A2-Level Biology

A2 Biology is seriously tricky — no question about that.
To do well, you're going to need to revise properly and practise hard.

This book has thorough notes on all the theory you need,
and it's got practice questions... lots of them.
For every topic there are warm-up and exam-style questions.

And of course, we've done our best to make the whole thing vaguely entertaining for you.

Complete Revision and Practice

Exam Board: OCR

Published by CGP

Editors:
Ellen Bowness, Katie Braid, Joe Brazier, Charlotte Burrows, Katherine Craig,
Rosie Gillham, Murray Hamilton, Jane Towle.

Contributors:
Gloria Barnett, Jessica Egan, Mark Ellingham, James Foster, Julian Hardwick, Derek Harvey,
Adrian Schmit, Sophie Watkins.

Proofreader:
Glenn Rogers.

ISBN: 978 1 84762 263 1

Groovy website: www.cgpbooks.co.uk
Jolly bits of clipart from CorelDRAW®
Printed by Elanders Ltd, Newcastle upon Tyne.

Based on the classic CGP style created by Richard Parsons.

Contents

The Scientific Process

This stuff may look similar to what you learnt at AS, but that's because you need to understand How Science Works for A2 as well. 'How Science Works' is all about the scientific process — how we develop and test scientific ideas. It's what scientists do all day, every day (well, except at coffee time — never come between a scientist and their coffee).

Scientists Come Up with Theories — Then Test Them...

Science tries to explain **how** and **why** things happen — it **answers questions**. It's all about seeking and gaining **knowledge** about the world around us. Scientists do this by **asking** questions and **suggesting** answers and then **testing** them, to see if they're correct — this is the **scientific process**.

1) **Ask** a question — make an **observation** and ask **why or how** it happens. E.g. why do plants grow faster in glasshouses than outside?
2) **Suggest** an answer, or part of an answer, by forming a **theory** (a possible **explanation** of the observations), e.g. glasshouses are warmer than outside and plants grow faster when it's warmer because the rate of photosynthesis is higher. (Scientists also sometimes form a **model** too — a **simplified picture** of what's physically going on.)
3) Make a **prediction** or **hypothesis** — a **specific testable statement**, based on the theory, about what will happen in a test situation. E.g. the rate of photosynthesis will be faster at 20 °C than at 10 °C.
4) Carry out a **test** — to provide **evidence** that will support the prediction (or help to disprove it). E.g. measure the rate of photosynthesis at various temperatures.

Simone predicted her hair would be worse on date night, based on the theory of sod's law.

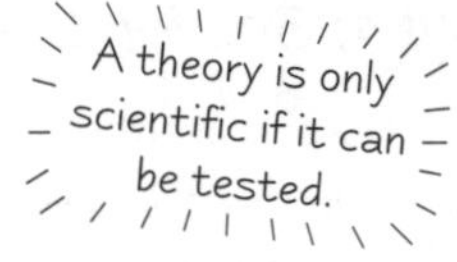

...Then They Tell Everyone About Their Results...

The results are **published** — scientists need to let others know about their work. Scientists publish their results in **scientific journals**. These are just like normal magazines, only they contain **scientific reports** (called papers) instead of the latest celebrity gossip.

1) Scientific reports are similar to the **lab write-ups** you do in school. And just as a lab write-up is **reviewed** (marked) by your teacher, reports in scientific journals undergo **peer review** before they're published.
2) The report is sent out to **peers** — other scientists who are experts in the **same area**. They examine the data and results, and if they think that the conclusion is reasonable it's **published**. This makes sure that work published in scientific journals is of a **good standard**.
3) But peer review **can't guarantee** the science is **correct** — other scientists still need to **reproduce** it.
4) Sometimes **mistakes** are made and flawed work is published. Peer review **isn't perfect** but it's probably the best way for scientists to self-regulate their work and to publish **quality reports**.

...Then Other Scientists Will Test the Theory Too

Other scientists read the published theories and results, and try to **test the theory** themselves. This involves:

- Repeating the **exact same experiments**.
- Using the theory to make **new predictions** and then testing them with **new experiments**.

If the Evidence Supports a Theory, It's Accepted — for Now

1) If all the experiments in all the world provide good evidence to back it up, the theory is thought of as **scientific 'fact'** (for now).
2) But it will never become **totally indisputable** fact. Scientific **breakthroughs or advances** could provide new ways to question and test the theory, which could lead to **new evidence** that **conflicts** with the current evidence. Then the testing starts all over again...

And this, my friend, is the **tentative nature of scientific knowledge** — it's always **changing** and **evolving**.

The Scientific Process

So scientists need evidence to back up their theories. They get it by carrying out experiments, and when that's not possible they carry out studies. But why bother with science at all? We want to know as much as possible so we can use it to try and improve our lives (and because we're nosy).

Evidence Comes from Lab Experiments...

1) Results from **controlled experiments** in **laboratories** are **great**.
2) A lab is the easiest place to **control variables** so that they're all **kept constant** (except for the one you're investigating).
3) This means you can draw meaningful **conclusions**.

For example, if you're investigating how light intensity affects the rate of photosynthesis you need to keep everything but the light intensity constant, e.g. the temperature, the concentration of carbon dioxide etc.

...and Well-Designed Studies

1) There are things you **can't** investigate in a lab, e.g. whether using a pesticide on farmland affects the number of non-pest species. You have to do a study instead.
2) You still need to try and make the study as controlled as possible to make it **more reliable**. But in reality it's **very hard** to control **all the variables** that **might** be having an effect.
3) You can do things to help, like having a **control** — e.g. an area of similar farmland nearby where the pesticide isn't applied. But you can't easily rule out every possibility.

Having a control reduced the effect of exercise on the study.

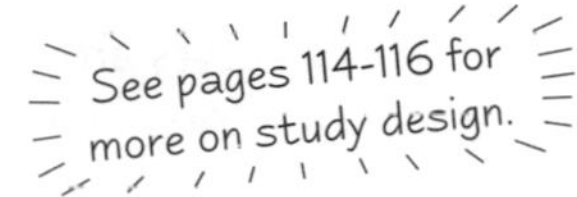
See pages 114-116 for more on study design.

Society Makes Decisions Based on Scientific Evidence

1) Lots of scientific work eventually leads to **important discoveries** or breakthroughs that could **benefit humankind**.
2) These results are **used by society** (that's you, me and everyone else) to **make decisions** — about the way we live, what we eat, what we drive, etc.
3) All sections of society use scientific evidence to make decisions, e.g. politicians use it to devise policies and individuals use science to make decisions about their own lives.

Other factors can **influence** decisions about science or the way science is used:

Economic factors

- Society has to consider the **cost** of implementing changes based on scientific conclusions — e.g. the **NHS** can't afford the most expensive drugs without **sacrificing** something else.
- Scientific research is **expensive** so companies won't always develop new ideas — e.g. developing new drugs is costly, so pharmaceutical companies often only invest in drugs that are likely to make them **money**.

Social factors

- **Decisions** affect **people's lives** — E.g. scientists may suggest **banning smoking** and **alcohol** to prevent health problems, but shouldn't **we** be able to **choose** whether **we** want to smoke and drink or not?

Environmental factors

- Scientists believe **unexplored regions** like remote parts of rainforests might contain **untapped drug** resources. But some people think we shouldn't **exploit** these regions because any interesting finds may lead to **deforestation** and **reduced biodiversity** in these areas.

So there you have it — how science works...

Hopefully these pages have given you a nice intro to how science works, e.g. what scientists do to provide you with 'facts'. You need to understand this, as you're expected to know how science works — for the exam and for life.

Communication and Homeostasis Basics

Ah, there's nothing like a nice long section to start you off on a new book — welcome to A2 Biology.

Responding to their Environment Helps Organisms Survive

1) **Animals increase** their **chances** of **survival** by **responding** to **changes** in their **external environment**, e.g. by **avoiding harmful environments** such as places that are too hot or too cold.
2) They also **respond** to **changes** in their **internal environment** to make sure that the **conditions** are always **optimal** for their **metabolism** (all the chemical reactions that go on inside them).
3) **Plants** also **increase** their **chances** of **survival** by **responding** to **changes** in their **environment** (see p. 98).
4) Any **change** in the internal or external **environment** is called a **stimulus**.

Receptors Detect Stimuli and Effectors Produce a Response

1) **Receptors detect stimuli.**
2) Receptors are **specific** — they only **detect one particular stimulus**, e.g. light, pressure or glucose concentration.
3) There are **many different types** of receptor that each detect a **different type of stimulus**.
4) Some receptors are **cells**, e.g. photoreceptors are receptor cells that connect to the nervous system. Some receptors are **proteins** on **cell surface membranes**, e.g. glucose receptors are proteins found in the cell membranes of some pancreatic cells.
5) **Effectors** are cells that bring about a **response** to a **stimulus**, to produce an **effect**. Effectors include **muscle cells** and cells found in **glands**, e.g. the **pancreas**.

Receptors Communicate with Effectors Via Hormones and Nerves

1) Receptors **communicate** with effectors via the **nervous system** (see p. 6) or the **hormonal system** (see p. 12), or sometimes using **both**.
2) **Nervous** and **hormonal communication** are both **examples** of **cell signalling** (ways cells communicate with each other).

Homeostasis is the Maintenance of a Constant Internal Environment

1) **Changes** in your **external environment** can affect your **internal environment** — the blood and tissue fluid that surrounds your cells.
2) **Homeostasis** involves **control systems** that keep your **internal environment** roughly **constant** (within **certain limits**).
3) **Keeping** your internal environment **constant** is vital for cells to **function normally** and to **stop** them being **damaged**.
4) It's particularly important to **maintain** the right **core body temperature**. This is because temperature affects **enzyme activity**, and enzymes **control** the **rate** of **metabolic reactions**:
   - If **body temperature** is **too high** (e.g. 40 °C) **enzymes** may become **denatured**. The enzyme's molecules **vibrate too much**, which **breaks** the **hydrogen bonds** that hold them in their **3D shape**. The **shape** of the enzyme's **active site** is **changed** and it **no longer works** as a **catalyst**. This means **metabolic reactions** are **less efficient**.
   - If body temperature is **too low enzyme activity** is **reduced**, **slowing** the rate of **metabolic reactions**.
   - The **highest rate** of **enzyme activity** happens at their **optimum temperature** (about **37 °C** in humans).
5) It's also important to **maintain** the right **concentration** of **glucose** in the **blood**, so there's always enough available for respiration.

There's more about control of body temperature on p. 14 and control of blood glucose on p. 16.

Communication and Homeostasis Basics

Homeostatic Systems Detect a Change and Respond by Negative Feedback

1) Homeostatic systems involve **receptors**, a **communication system** and **effectors** (see the previous page).
2) Receptors detect when a level is **too high** or **too low**, and the information's communicated via the **nervous** system or the **hormonal** system to **effectors**.
3) The effectors respond to **counteract** the change — bringing the level **back** to **normal**.
4) The mechanism that **restores** the level to **normal** is called a **negative feedback** mechanism.
5) Negative feedback **keeps** things around the **normal** level, e.g. body temperature is usually kept **within 0.5 °C** above or below **37 °C**.
6) Negative feedback only works within **certain limits** though — if the change is **too big** then the **effectors** may **not** be able to **counteract** it, e.g. a huge drop in body temperature caused by prolonged exposure to cold weather may be too large to counteract.

Control of **body temperature** by **negative feedback**:

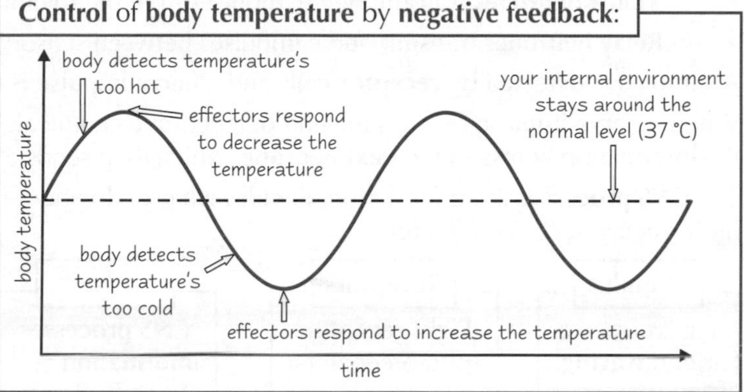

Positive Feedback Mechanisms Amplify a Change from the Normal Level

1) Some changes trigger a **positive feedback** mechanism, which **amplifies** the change.
2) The effectors respond to **further increase** the level **away** from the **normal** level.
3) Positive feedback is useful to **rapidly activate** something, e.g. a **blood clot** after an injury:
   - **Platelets** become **activated** and release a **chemical** — this triggers **more platelets** to be activated, and so on.
   - Platelets **very quickly** form a **blood clot** at the injury site.
   - The process **ends** with **negative feedback**, when the body detects the **blood clot** has been **formed**.
4) Positive feedback **isn't** involved in **homeostasis** because it **doesn't** keep your internal environment **constant**.

Practice Questions

Q1 Why do organisms respond to changes in their environment?

Q2 What is a stimulus?

Q3 Give two types of effector.

Q4 What is cell signalling?

Q5 What is a negative feedback mechanism?

Q6 What type of mechanism amplifies a change from the normal level?

Exam Question

Q1 a) Define homeostasis. [1 mark]

b) Describe the role of receptors, communication systems and effectors in homeostasis. [3 marks]

Responding to questions in an exam helps you to pass...

Animals respond to changes in their internal and external environment. They respond to internal changes so that they can keep conditions just right for all their bodily reactions. Maintaining this constant environment is called homeostasis — basically you just need to remember that if one thing goes up the body responds to bring it down, and vice versa.

The Nervous System and Neurones

The nervous system helps organisms to respond to the environment, so you need to know a bit more about it...

The ***Nervous System*** *Sends Information as* ***Nerve Impulses***

1) The **nervous system** is made up of a **complex network** of cells called **neurones**. There are **three main types** of neurone:
   - **Sensory neurones** transmit nerve impulses from **receptors** to the **central nervous system** (**CNS**) — the **brain** and **spinal cord**.
   - **Motor neurones** transmit nerve impulses from the **CNS** to **effectors**.
   - **Relay neurones** transmit nerve impulses **between** sensory neurones and motor neurones.
2) A stimulus is detected by **receptor cells** and a **nerve impulse** is sent along a **sensory neurone**.
3) When a **nerve impulse** reaches the end of a neurone chemicals called **neurotransmitters** take the information across to the **next neurone**, which then sends a **nerve impulse** (see p. 10).
4) The **CNS processes** the information, **decides what to do** about it and sends impulses along **motor neurones** to an **effector**.

Nerve impulses are electrical impulses. They're also called action potentials.

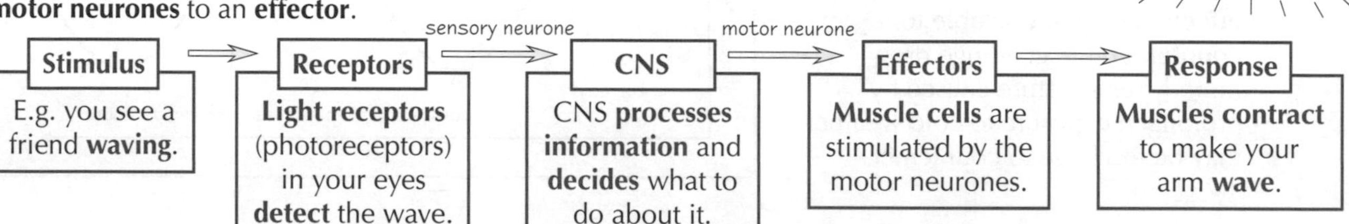

Sensory Receptors *Convert* ***Stimulus Energy*** *into* ***Nerve Impulses***

1) **Different stimuli** have **different forms** of **energy**, e.g. light energy or chemical energy.
2) But your **nervous system** only sends information in the form of **nerve impulses** (electrical impulses).
3) **Sensory receptors convert** the energy of a **stimulus** into **electrical energy**.
4) So, sensory receptors act as **transducers** — something that **converts** one form of energy into another.
5) Here's a bit more about how receptor cells that communicate information via the **nervous system** work:

- When a nervous system receptor is in its **resting state** (not being stimulated), there's a **difference in charge** between the **inside** and the **outside** of the cell — this is generated by ion pumps and ion channels. This means there's a **voltage** across the membrane. Voltage is also known as **potential difference**.
- The **potential difference** when a cell is at **rest** is called its **resting potential**. When a stimulus is detected, the cell membrane is **excited** and becomes **more permeable**, allowing **more ions** to move **in and out** of the cell — **altering** the **potential difference**. The **change** in **potential difference** due to a stimulus is called the **generator potential**.
- A **bigger stimulus** excites the membrane more, causing a **bigger movement** of ions and a **bigger change** in potential difference — so a **bigger generator potential** is produced.
- If the **generator potential** is **big enough** it'll trigger an **action potential** (nerve impulse) along a neurone. An action potential is only triggered if the generator potential reaches a certain level called the **threshold** level.
- If the stimulus is **too weak** the generator potential **won't reach** the **threshold**, so there's **no action potential**.

You Need to ***Learn*** *the* ***Structure*** *of* ***Sensory Neurones...***

1) All neurones have a **cell body** with a **nucleus** (plus **cytoplasm** and all the other **organelles** you usually get in a cell).
2) The cell body has **extensions** that **connect** to **other neurones** — **dendrites** and **dendrons** carry nerve impulses **towards** the **cell body**, and **axons** carry nerve impulses **away** from the **cell body**.
3) Sensory neurones have **short dendrites** and **one long dendron** to carry nerve impulses from **receptor cells** to the **cell body**, and **one short axon** that carries impulses from the **cell body** to the **CNS**.

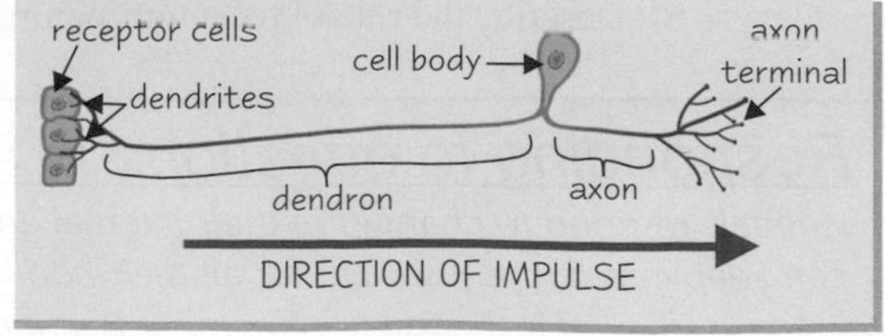

The Nervous System and Neurones

...and Motor Neurones

Motor neurones have **many short dendrites** that carry nerve impulses from the **central nervous system** (**CNS**) to the **cell body**, and **one long axon** that carries nerve impulses from the **cell body** to **effector cells**.

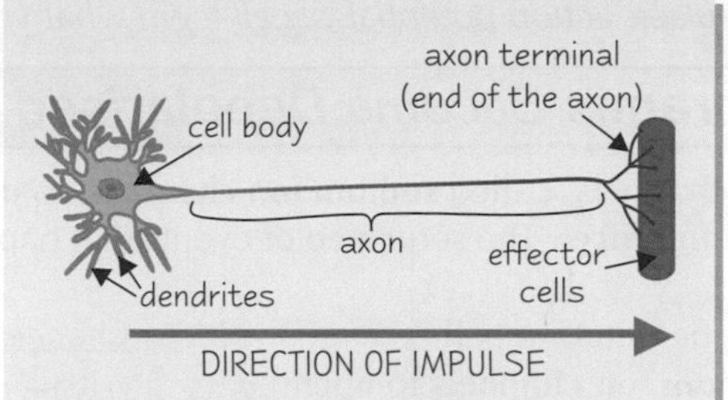

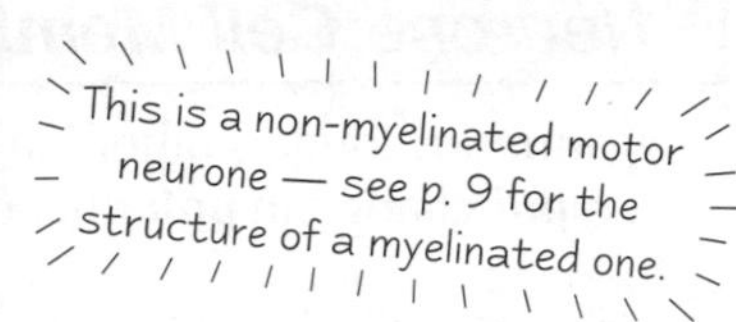

Neurone *Cell Membranes* are *Polarised* at *Rest*

1) In a neurone's **resting state** (when it's not being stimulated), the **outside** of the membrane is **positively charged** compared to the **inside**. This is because there are **more positive ions outside** the cell than inside.
2) So the membrane is **polarised** — there's a **difference in charge**.
3) The voltage across the membrane when it's at rest is called the **resting potential** — it's about **–70 mV**.
4) The resting potential is created and maintained by **sodium-potassium pumps** and **potassium ion channels** in a neurone's membrane:

Sodium-potassium pump

These pumps use **active transport** to move **three sodium ions** (Na^+) **out** of the neurone for every **two potassium ions** (K^+) moved **in**. **ATP** is needed to do this.

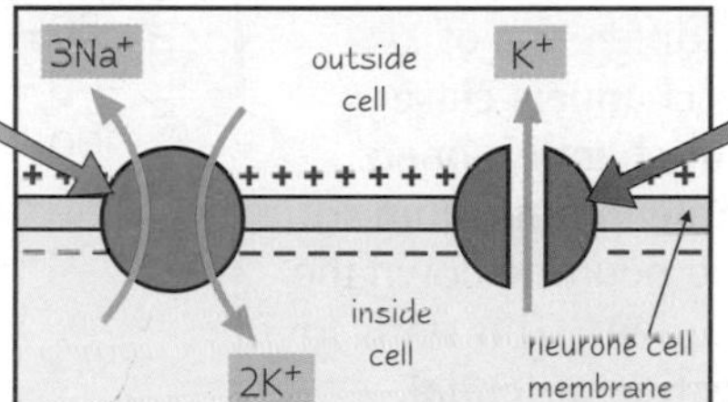

Potassium ion channel

These channels allow **facilitated diffusion** of **potassium ions** (K^+) **out** of the neurone, down their **concentration gradient**.

- The sodium-potassium pumps move **sodium ions out** of the neurone, but the membrane **isn't permeable** to **sodium ions**, so they **can't diffuse back in**. This creates a **sodium ion electrochemical gradient** (a **concentration gradient** of **ions**) because there are **more** positive sodium ions **outside** the cell than inside.
- The sodium potassium pumps also move **potassium ions in** to the neurone, but the membrane **is permeable** to **potassium ions** so they **diffuse back out** through potassium ion channels.
- This makes the **outside** of the cell **positively charged** compared to the inside.

Practice Questions

Q1 What is the function of a motor neurone?

Q2 What do sensory receptors convert energy into?

Q3 Name the pumps and channels that maintain a neurone's resting potential.

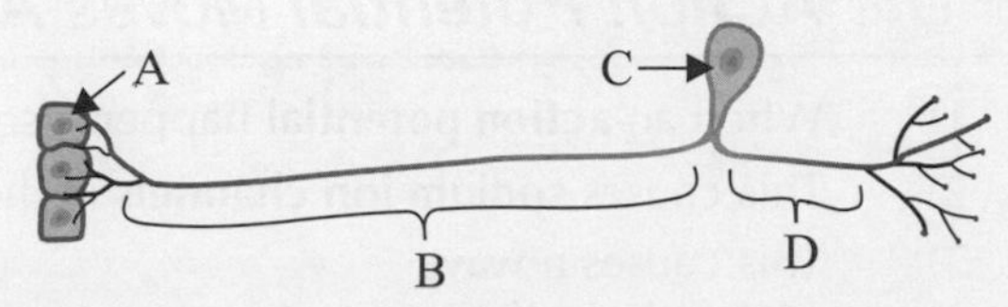

Exam Question

Q1 Bright light causes circular iris muscles in an animal's eyes to contract, which constricts the pupils and protects the eyes. Describe and explain the roles of receptors and effectors for this response. [5 marks]

Q2 The diagram above is of a sensory neurone. Name parts A to D. [4 marks]

Vacancy — talented gag writer required for boring biology topics...

Actually, it's not that boring, it's just all the stuff about sensory receptors and resting potentials can be a bit tricky to get your head around. Just take your time and try scribbling it all down a few times till it starts to make some kind of sense. Then you can finish off by drawing loads of sensory and motor neurones, until you can label them in your sleep.

Action Potentials

Electrical impulses, nerve impulses, action potentials... call them what you will, you need to know how they work.

Neurone **Cell Membranes** *Become* **Depolarised** *when they're* **Stimulated**

A **stimulus** triggers other ion channels, called **sodium ion channels**, to **open**. If the stimulus is big enough, it'll trigger a **rapid change** in **potential difference**. The sequence of events that happen are known as an **action potential**:

① **Stimulus** — this **excites** the neurone cell membrane, causing **sodium ion channels** to **open**. The membrane becomes **more permeable** to sodium, so **sodium ions diffuse into** the neurone down the sodium ion electrochemical gradient. This makes the **inside** of the neurone **less negative**.

② **Depolarisation** — if the potential difference reaches the **threshold** (around **–55 mV**), **voltage-gated sodium ion channels open**. **More sodium ions diffuse into** the neurone.

Voltage-gated ion channels open at a certain voltage.

③ **Repolarisation** — at a potential difference of around **+30 mV** the **sodium ion channels close** and **voltage-gated potassium ion channels open**. The membrane is **more permeable** to potassium so **potassium ions diffuse out** of the neurone down the potassium ion concentration gradient. This starts to get the membrane **back** to its **resting potential**.

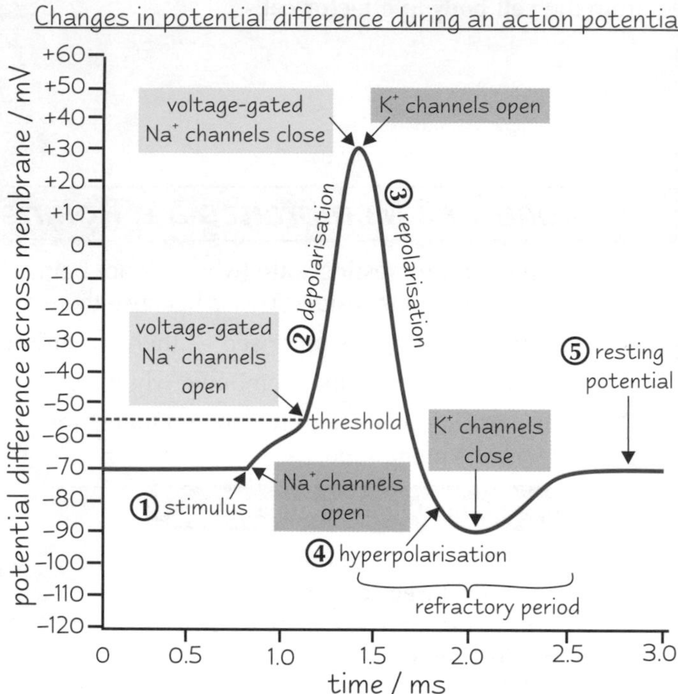

④ **Hyperpolarisation** — **potassium ion channels** are **slow to close** so there's a slight '**overshoot**' where too many potassium ions diffuse out of the neurone. The potential difference becomes **more negative** than the **resting potential** (i.e. less than –70 mV).

⑤ **Resting potential** — the ion channels are **reset**. The **sodium-potassium pump** returns the membrane to its **resting potential** and maintains it until the membrane's excited by another stimulus.

After an **action potential**, the neurone cell membrane **can't** be **excited** again straight away. This is because the ion channels are **recovering** and they **can't** be made to **open** — **sodium ion channels** are **closed** during repolarisation and **potassium ion channels** are **closed** during hyperpolarisation. This period of recovery is called the **refractory period**.

The **Action Potential** *Moves* **Along** *the* **Neurone** *as a* **Wave** *of* **Depolarisation**

1) When an **action potential** happens, some of the **sodium ions** that enter the neurone **diffuse sideways**.
2) This causes **sodium ion channels** in the **next region** of the neurone to **open** and **sodium ions diffuse into** that part.
3) This causes a **wave of depolarisation** to travel along the neurone.
4) The **wave** moves **away** from the parts of the membrane in the **refractory period** because these parts **can't fire** an action potential.

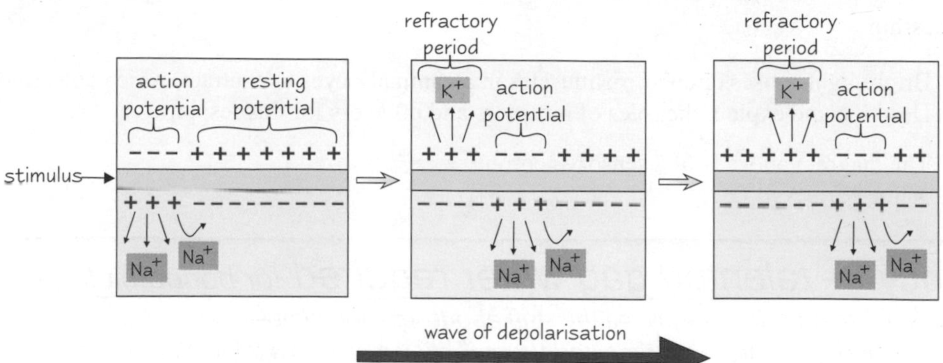

Action Potentials

A Bigger Stimulus Causes More Frequent Impulses

1) Once the threshold is reached, an action potential will **always fire** with the **same change in voltage**, no matter how big the stimulus is.
2) If **threshold isn't reached**, an action potential **won't fire**.
3) A **bigger stimulus** won't cause a bigger action potential, but it will cause them to fire **more frequently**.

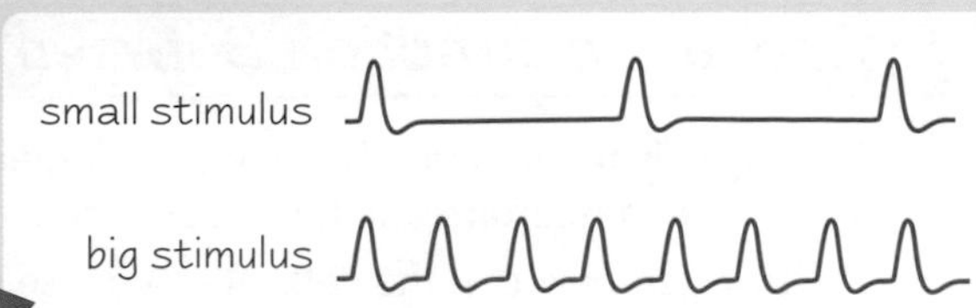

Action Potentials Go Faster in Myelinated Neurones

1) Some neurones are **myelinated** — they have a **myelin sheath**.
2) The myelin sheath is an **electrical insulator**.
3) It's made of a type of cell called a **Schwann cell**.
4) Between the Schwann cells are tiny patches of **bare membrane** called the **nodes of Ranvier**. **Sodium ion channels** are **concentrated** at the nodes.
5) In a **myelinated** neurone, **depolarisation** only happens at the **nodes of Ranvier** (where sodium ions can get through the membrane).
6) The neurone's **cytoplasm conducts** enough electrical charge to **depolarise** the **next node**, so the impulse '**jumps**' from node to node.
7) This is called **saltatory conduction** and it's **really fast**.
8) In a **non-myelinated** neurone, the impulse travels as a **wave** along the **whole length** of the **axon membrane**.
9) This is **slower** than saltatory conduction (although it's still pretty quick).

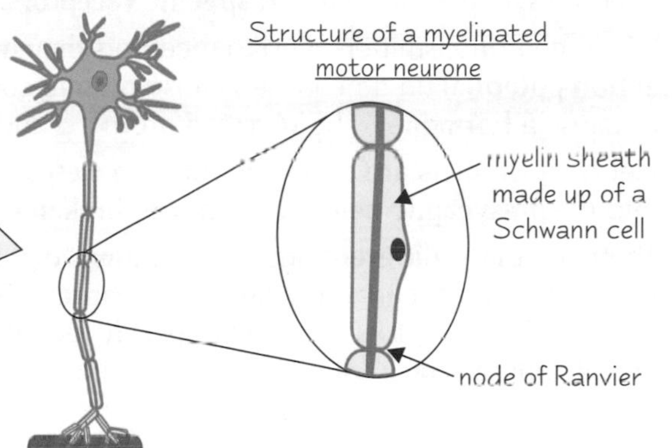

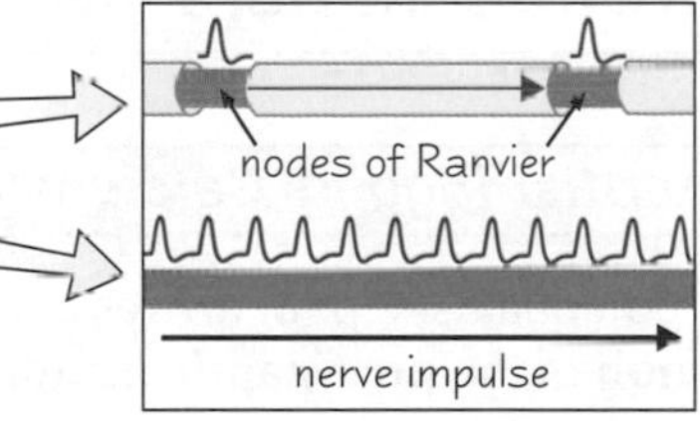

Practice Questions

Q1 Briefly describe how an action potential moves along a neurone.

Q2 What are nodes of Ranvier?

Exam Questions

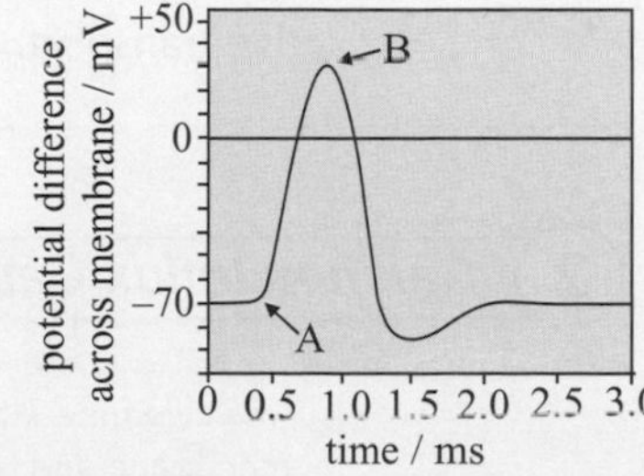

Q1 The graph shows an action potential across an axon membrane following the application of a stimulus.

a) What label should be added at point A? [1 mark]

b) Explain what causes the change in potential difference between point A and point B. [3 marks]

c) A stimulus was applied at 1.5 ms, but failed to produce an action potential. Suggest why. [2 marks]

Q2 Multiple sclerosis is a disease of the nervous system characterised by damage to the myelin sheaths of neurones. Explain how this will affect the transmission of action potentials. [5 marks]

I'm feeling a bit depolarised after all that...

Action potentials are potentially confusing. The way I remember them is that polarisation is the difference in charge across the cell's membrane — during depolarisation that difference becomes smaller and during repolarisation it gets bigger again.

Synapses

When an action potential arrives at the end of a neurone the information has to be passed on to the next cell — this could be another neurone, a muscle cell or a gland cell.

*A **Synapse** is a **Junction** Between a **Neurone** and the **Next Cell***

1) A **synapse** is the junction between a **neurone** and another **neurone**, or between a **neurone** and an **effector cell**, e.g. a muscle or gland cell.
2) The **tiny gap** between the cells at a synapse is called the **synaptic cleft**.
3) The **presynaptic neurone** (the one before the synapse) has a **swelling** called a **synaptic knob**. This contains **synaptic vesicles** filled with **chemicals** called **neurotransmitters**.
4) When an **action potential** reaches the end of a neurone it causes **neurotransmitters** to be **released** into the synaptic cleft. They **diffuse across** to the **postsynaptic membrane** (the one after the synapse) and **bind** to **specific receptors**.
5) When neurotransmitters bind to receptors they might **trigger** an **action potential** (in a neurone), cause **muscle contraction** (in a muscle cell), or cause a **hormone** to be **secreted** (from a gland cell).
6) Neurotransmitters are **removed** from the **cleft** so the **response** doesn't keep happening, e.g. they're taken back into the **presynaptic neurone** or they're **broken down** by **enzymes** (and the products are taken into the neurone).
7) There are many **different** neurotransmitters, e.g. **acetylcholine** (**ACh**) and **noradrenaline**. Synapses that use acetylcholine are called **cholinergic synapses**. Their structure is exactly the **same** as in the diagram above. They bind to receptors called **cholinergic receptors**, and they're broken down by an enzyme called **acetylcholinesterase** (**AChE**).

Typical structure of a synapse

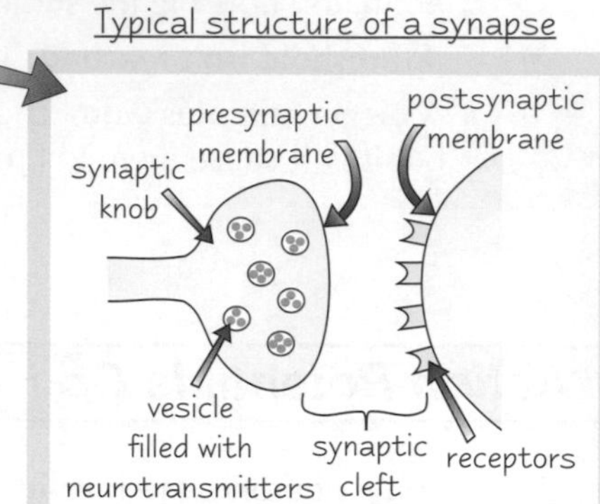

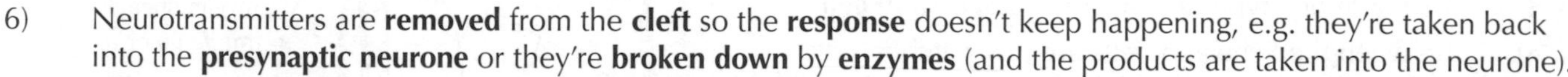

*Here's How **Neurotransmitters Transmit Nerve Impulses Between Neurones***

1 *An **Action Potential** Triggers **Calcium Influx***

1) An action potential (see p. 8) arrives at the **synaptic knob** of the **presynaptic neurone**.
2) The action potential stimulates **voltage-gated calcium ion channels** in the **presynaptic neurone** to **open**.
3) **Calcium ions diffuse into** the synaptic knob. (They're pumped out afterwards by active transport.)

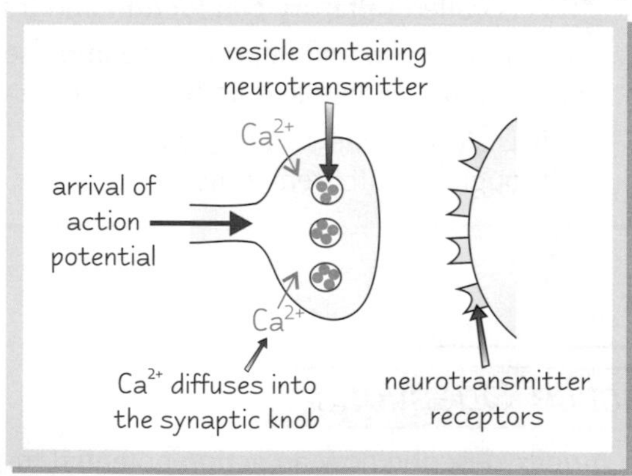

2 ***Calcium Influx** Causes **Neurotransmitter Release***

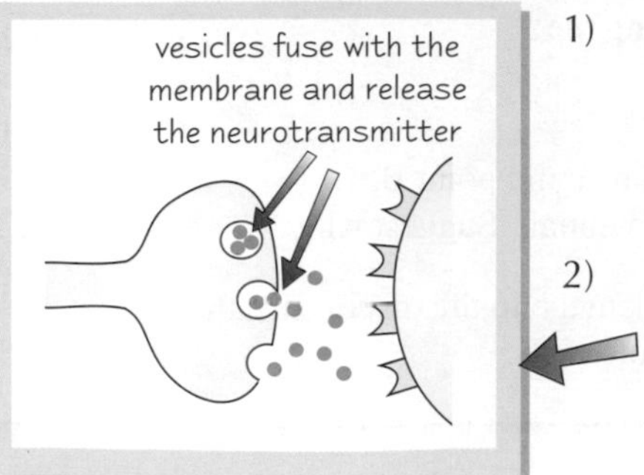

1) The influx of **calcium ions** into the synaptic knob causes the **synaptic vesicles** to **move** to the **presynaptic membrane**. They then **fuse** with the presynaptic membrane.
2) The **vesicles release** the neurotransmitter into the **synaptic cleft** — this is called **exocytosis**.

Synapses

3) The Neurotransmitter Triggers an Action Potential in the Postsynaptic Neurone

1) The neurotransmitter **diffuses** across the **synaptic cleft** and **binds** to specific **receptors** on the **postsynaptic membrane**.
2) This causes **sodium ion channels** in the **postsynaptic neurone** to **open**.
3) The **influx** of **sodium ions** into the postsynaptic membrane causes **depolarisation**. An **action potential** on the **postsynaptic membrane** is generated if the **threshold** is reached.
4) The **neurotransmitter** is **removed** from the **synaptic cleft** so the **response** doesn't keep happening.

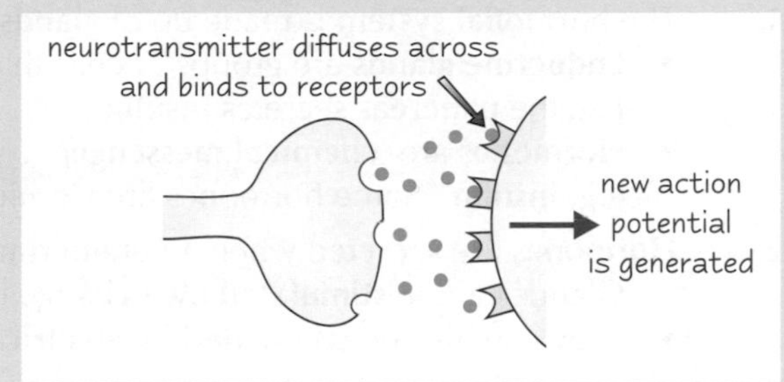

Synapses Play Vital Roles in the Nervous System

1) Synapses allow Information to be Dispersed or Amplified

1) When **one** neurone **connects** to **many** neurones information can be **dispersed** to **different parts** of the body. This is called **synaptic divergence**.
2) When **many** neurones **connect** to **one** neurone information can be **amplified** (made stronger). This is called **synaptic convergence**.

2) Summation at Synapses Finely Tunes the Nervous Response

If a stimulus is **weak**, only a **small amount** of **neurotransmitter** will be released from a neurone into the synaptic cleft. This might not be enough to **excite** the postsynaptic membrane to the **threshold** level and stimulate an action potential. **Summation** is where the effect of neurotransmitter released from **many neurones** (or **one** neurone that's stimulated **a lot** in a short period of time) is **added together**.

3) Synapses make sure Impulses are Transmitted One Way

Receptors for neurotransmitters are **only** on the **postsynaptic** membranes, so synapses make sure **impulses** can only travel in **one direction**.

Practice Questions

Q1 Give one way that neurotransmitters are removed from the synaptic cleft.

Q2 What neurotransmitter do you find at cholinergic synapses?

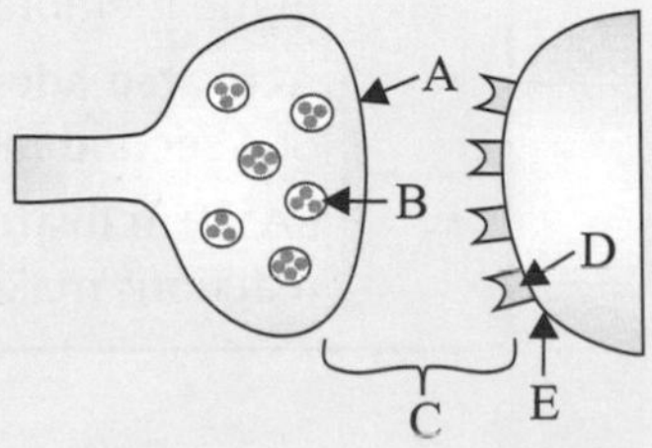

Exam Questions

Q1 The diagram on the right shows a synapse. Label parts A-E. [5 marks]

A
B
C
D
E

Q2 Describe the sequence of events from the arrival of an action potential at the presynaptic membrane of a synapse to the generation of a new action potential at the postsynaptic membrane. [6 marks]

Synaptic knobs and clefts — will you stop giggling at the back...

Some more pretty tough pages here, aren't I kind to you. And lots more diagrams to have a go at drawing and re-drawing. Don't worry if you're not the world's best artist, just make sure you add labels to your drawings to explain what's happening.

The Hormonal System and Glands

Now you've seen how the nervous system helps us respond to our environment, it's on to the hormonal system...

*The **Hormonal System** Sends Information as **Chemical Signals***

1) The **hormonal system** is made up of **glands** (called **endocrine glands**) and **hormones**:
   - **Endocrine glands** are groups of cells that are specialised to **secrete hormones.** E.g. the **pancreas** secretes **insulin**.
   - **Hormones** are '**chemical messengers**'. Many hormones are **proteins** or **peptides**, e.g. **insulin**. Some hormones are **steroids**, e.g. **progesterone**.

The hormonal system is also called the endocrine system.

2) **Hormones** are **secreted** when an **endocrine gland** is **stimulated**:
   - Glands can be **stimulated** by a **change** in **concentration** of a specific **substance** (sometimes **another hormone**).
   - They can also be **stimulated** by **electrical impulses**.
3) Hormones **diffuse directly into** the **blood**, then they're **taken** around the body by the **circulatory system**.
4) They **diffuse out** of the blood **all over** the **body** but each hormone will only **bind** to **specific receptors** for that hormone, found on the membranes of some cells, called **target cells**. Tissue that contains target cells is called **target tissue**.
5) The hormones trigger a **response** in the **target cells** (the **effectors**).

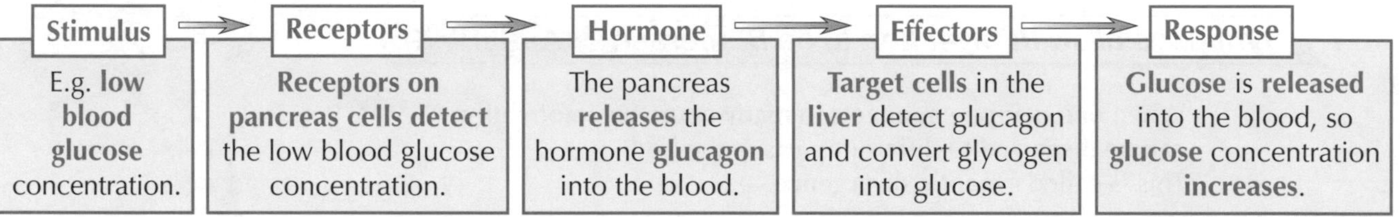

Hormones** Bind to **Receptors** and Cause a **Response** via **Second Messengers

1) A **hormone** is called a **first messenger** because it carries the chemical message the **first part** of the way, from the **endocrine gland** to the **receptor** on the **target cells**.
2) When a hormone **binds** to its receptor it **activates** an **enzyme** in the **cell membrane**.
3) The enzyme catalyses the **production** of a **molecule** inside the cell called a **signalling molecule** — this molecule **signals** to **other parts** of the cell to **change** how the cell **works**.
4) The **signalling molecule** is called a **second messenger** because it carries the chemical message the **second part** of the way, from the **receptor** to **other parts** of the **cell**.
5) Second messengers **activate** a **cascade** (a chain of reactions) **inside** the cell. Here's an **example** you need to **learn**:

- The hormone **adrenaline** is a **first messenger**.
- It binds to **specific receptors** in the **cell membranes** of many cells, e.g. liver cells.
- When adrenaline binds it **activates** an **enzyme** in the membrane called **adenylate cyclase**.
- **Activated adenylate cyclase** catalyses the production of a **second messenger** called **cyclic AMP** (**cAMP**).
- cAMP **activates** a **cascade**, e.g. a cascade of enzyme reactions make **more glucose available** to the cell.

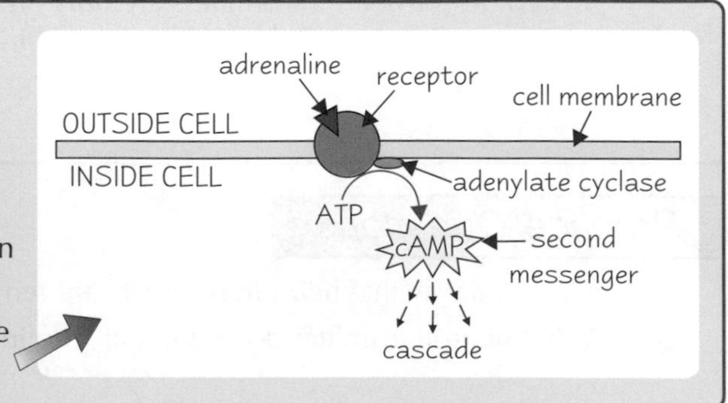

*There are **Two Types** of **Gland — Exocrine** and **Endocrine***

1) Exocrine glands secrete chemicals through **ducts** (tubes) into **cavities** or onto the **surface** of the body, e.g. **sweat glands** secrete **sweat** onto the **skin surface**.
2) They usually secrete **enzymes**, e.g. **digestive glands** secrete **digestive enzymes** into the **gut**.
3) **Endocrine** glands secrete **hormones directly** into the **blood**.
4) **Some organs** have exocrine tissue **and** endocrine tissue, so act as both types of gland.

The Hormonal System and Glands

*The **Pancreas** is an **Exocrine** and an **Endocrine Gland***

The pancreas is a gland that's found **below** the **stomach**.
You need to know about its exocrine function and its endocrine function:

Exocrine function of the pancreas

1) **Most** of the pancreas is **exocrine** tissue.
2) The exocrine cells are called **acinar cells**.
3) They're found in **clusters** around the **pancreatic duct** — a duct that goes to the **duodenum** (part of the small intestine).
4) The acinar cells **secrete digestive enzymes** into the **pancreatic duct**.
5) The enzymes **digest food** in the **duodenum**, e.g. **amylase** breaks down **starch** to **glucose**.

Endocrine function of the pancreas

1) The areas of **endocrine** tissue are called the **islets of Langerhans**.
2) They're found in clusters around **blood capillaries**.
3) The islets of Langerhans **secrete hormones** directly into the **blood**.
4) They're made up of **two types** of cell:
   - **Alpha** (α) **cells** secrete a **hormone** called **glucagon**.
   - **Beta** (β) **cells** secrete a **hormone** called **insulin**.
5) **Glucagon** and **insulin** help to **control blood glucose concentration** (see p. 16).

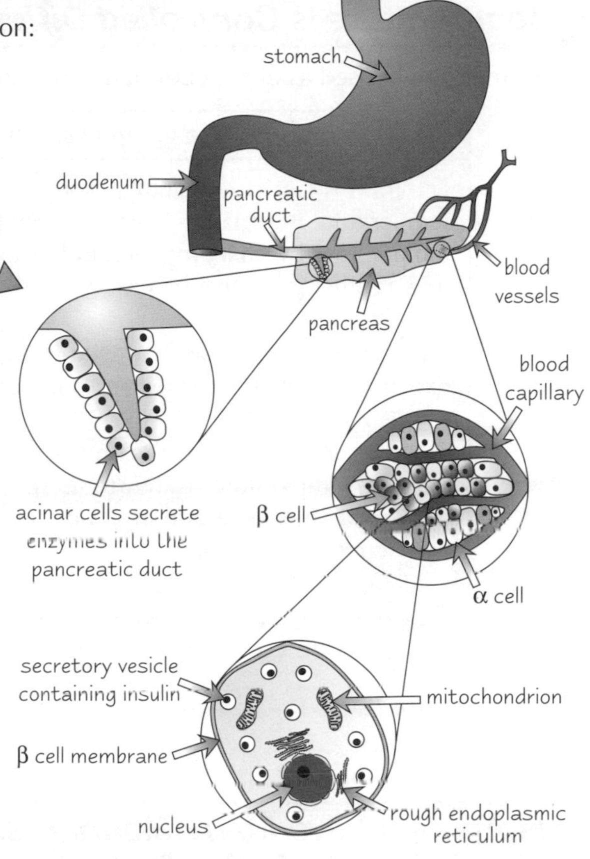

*The **Adrenal Glands** Secrete **Hormones** Including **Adrenaline***

1) The **adrenal glands** are **endocrine glands** that are found just **above** your **kidneys**.
2) Each adrenal gland has an **outer** part called the **cortex** and an **inner** part called the **medulla**.
3) The cortex and the medulla have **different functions**:

- The **cortex** secretes **steroid hormones**, e.g. it secretes **cortisol** when you're **stressed**.
- The **medulla** secretes **catecholamine hormones** (modified amino acids), e.g. it secretes **adrenaline** when you're **stressed**.

Cortisol and adrenaline work together to control your response to stress.

Practice Questions

Q1 What is a target tissue?

Q2 What two hormones do the islets of Langerhans in the pancreas secrete?

Q3 Give one function of the cortex of the adrenal gland.

Exam Questions

Q1 In the endocrine system, explain the difference between a first messenger and a second messenger. [4 marks]

Q2 Describe the differences between an endocrine gland and an exocrine gland. [2 marks]

Islets of Langerhans — sounds like an exotic beach to me...

All this talk of the "islets of Langerhans" and I can think of nothing else but sun, sea and sand... but it's secretions, second messengers and cyclic AMP for you, until your exams are over and you can start planning any holidays.

Homeostasis — Control of Body Temperature

Homeostasis is responsible for controlling body temperature in mammals like you — stopping you freezing or becoming a hot sweaty mess. Other organisms control their body temperature differently. Read on, oh chosen one, read on...

Temperature is Controlled Differently in Ectotherms and Endotherms

Animals are classed as either **ectotherms** or **endotherms**, depending on how they **control** their body temperature:

Ectotherms — e.g. **reptiles, fish**	**Endotherms** — e.g. **mammals, birds**
Ectotherms **can't control** their body temperature **internally** — they **control** their temperature by **changing** their **behaviour** (e.g. reptiles gain heat by basking in the sun).	Endotherms **control** their body temperature **internally** by **homeostasis.** They can also control their temperature by **behaviour** (e.g. by finding shade).
Their **internal** temperature **depends** on the **external temperature** (their surroundings).	Their internal temperature is **less affected** by the **external temperature** (within certain limits).
Their **activity** level **depends** on the external temperature — they're **more** active at **higher** temperatures and **less** active at **lower** temperatures.	Their **activity** level is largely **independent** of the **external temperature** — they can be active at any temperature (within certain limits).
They have a **variable metabolic rate** and they **generate** very **little heat** themselves.	They have a constantly **high metabolic rate** and they **generate** a **lot of heat** from metabolic reactions.

Mammals have Many Mechanisms to Change Body Temperature

Mechanisms to REDUCE body temperature:

Sweating — **more sweat** is secreted from **sweat glands** when the body's too hot. The water in sweat **evaporates** from the surface of the skin and **takes heat** from the body. The **skin** is **cooled**.

Hairs lie flat — mammals have a layer of **hair** that provides **insulation** by **trapping air** (air is a poor conductor of heat). When it's hot, **erector pili muscles relax** so the hairs lie flat. **Less air** is trapped, so the skin is **less insulated** and **heat** can be **lost** more easily.

Vasodilation — when it's hot, **arterioles** near the surface of the skin **dilate** (this is called **vasodilation**). **More blood** flows through the **capillaries** in the surface layers of the dermis. This means **more heat** is **lost** from the skin by **radiation** and the **temperature** is **lowered**.

Mechanisms to INCREASE body temperature:

Shivering — when it's cold, **muscles contract** in **spasms**. This makes the body **shiver** and **more heat** is **produced** from **increased respiration**.

Much less sweat — **less sweat** is secreted from sweat glands when it's cold, **reducing** the amount of **heat loss**.

Hairs stand up — **erector pili muscles contract** when it's cold, which makes the **hairs stand up**. This **traps more air** and so **prevents heat loss**.

Vasoconstriction — when it's cold, **arterioles** near the surface of the skin **constrict** (this is called **vasoconstriction**) so **less blood** flows through the **capillaries** in the surface layers of the dermis. This **reduces heat loss**.

Hormones — the body releases **adrenaline** and **thyroxine**. These **increase metabolism** and so **more heat is produced**.

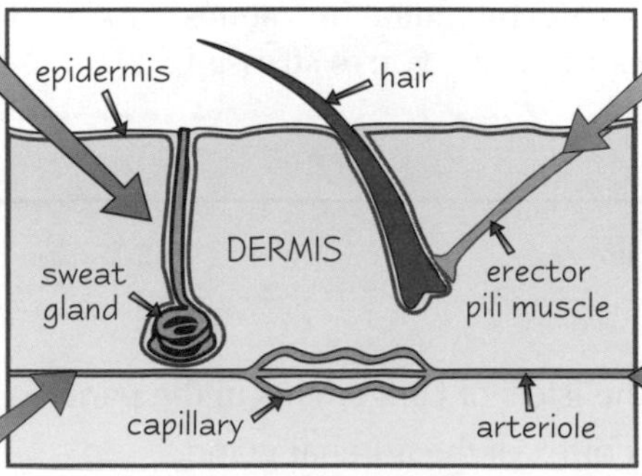

epidermis
hair
DERMIS
sweat gland
erector pili muscle
capillary
arteriole

Homeostasis — Control of Body Temperature

The **Hypothalamus Controls** Body Temperature in **Mammals**

1) **Body temperature** in mammals is **maintained** at a **constant level** by a part of the **brain** called the **hypothalamus**.
2) The hypothalamus **receives information** about **temperature** from **thermoreceptors** (temperature receptors):
   - Thermoreceptors in the **hypothalamus** detect **internal temperature** (the temperature of the blood).
   - Thermoreceptors in the **skin** (called **peripheral temperature receptors**) detect **external temperature** (the temperature of the skin).
3) Thermoreceptors send **impulses** along **sensory neurones** to the **hypothalamus**, which sends **impulses** along **motor neurones** to **effectors** (e.g. **skeletal muscles**, or **sweat glands** and **erector pili muscles** in the **skin**).
4) The effectors respond to **restore** the body temperature **back** to **normal**. Here's how it all works:

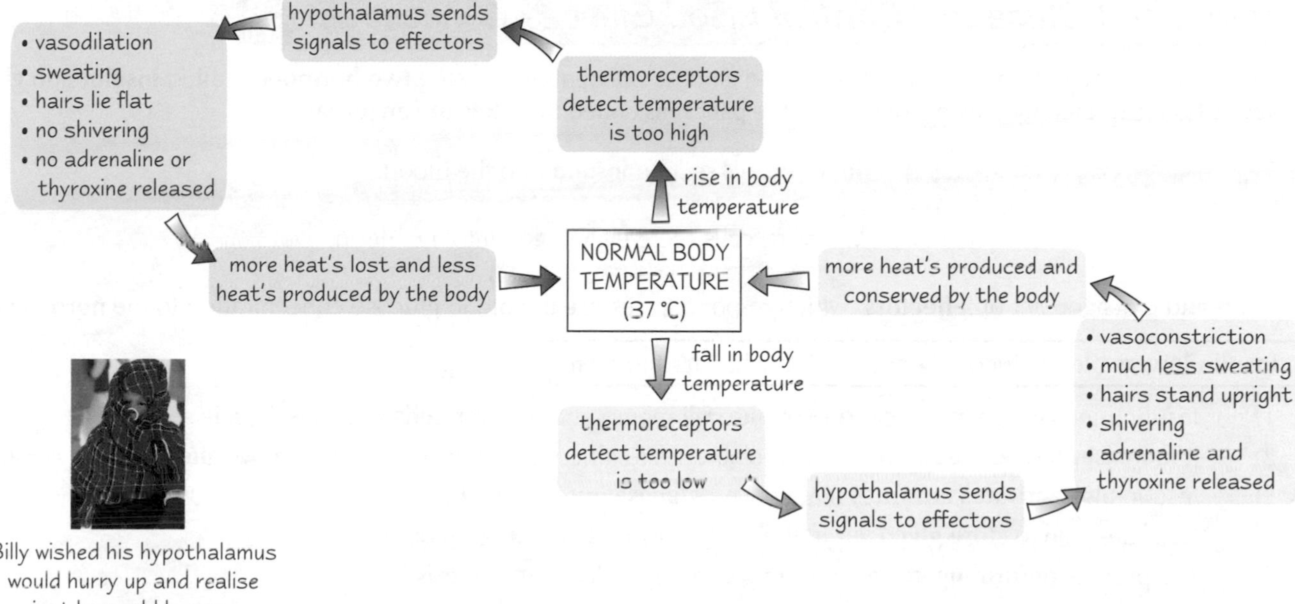

Billy wished his hypothalamus would hurry up and realise just how cold he was.

Practice Questions

Q1 Give four differences between ectotherms and endotherms.

Q2 Which type of animal has more control over their body temperature, ectotherms or endotherms?

Q3 How does sweating reduce body temperature?

Q4 How does vasodilation help the body to lose heat?

Q5 Which part of the brain is responsible for maintaining a constant body temperature in mammals?

Exam Questions

Q1 Snakes are usually found in warm climates. Suggest why they are not usually found in cold climates. Explain your answer. [4 marks]

Q2 Mammals that live in cold climates have thick fur and layers of fat beneath their skin to keep them warm. Describe and explain two other ways they maintain a constant body temperature in cold conditions. [4 marks]

Q3 Describe and explain how the body detects a high external temperature. [2 marks]

Sweat, hormones and erector muscles — ooooh errrrrrr...

The mechanisms that change body temperature are pretty good and can cope with some extreme temperatures, but I reckon I could think up some slightly less embarrassing ways of doing it, instead of getting all red-faced and stinky. Mind you, it seems like ectotherms have got it sussed with their whole sunbathing thing — now that's definitely the life...

Homeostasis — Control of Blood Glucose

These pages are all about how homeostasis helps you to not go totally hyper when you stuff your face with sweets.

Eating** and **Exercise** Change the **Concentration** of **Glucose** in your **Blood

1) **All cells** need a constant **energy supply** to work — so **blood glucose concentration** must be carefully **controlled**.
2) The **concentration** of **glucose** in the blood is **normally** around **90 mg per 100 cm³** of blood. It's **monitored** by cells in the **pancreas**.
3) Blood glucose concentration **rises** after **eating food** containing **carbohydrate**.
4) Blood glucose concentration **falls** after **exercise**, as **more glucose** is used in **respiration** to **release energy**.

***Insulin** and **Glucagon Control** Blood Glucose Concentration*

The hormonal system (see p. 12) **controls** blood glucose concentration using **two hormones** called **insulin** and **glucagon**. They're both **secreted** by clusters of cells in the **pancreas** called the **islets of Langerhans**:

Beta (β) cells secrete **insulin** into the blood.

Alpha (α) cells secrete **glucagon** into the blood.

Insulin and glucagon act on **effectors**, which respond to **restore** the blood glucose concentration to the **normal level**:

Insulin lowers blood glucose concentration when it's **too high**

1) Insulin binds to **specific receptors** on the cell membranes of **liver cells** and **muscle cells**.
2) It **increases** the **permeability** of cell membranes to glucose, so the cells **take up more glucose**.
3) Insulin also **activates enzymes** that convert **glucose** into **glycogen**.
4) Cells are able to **store glycogen** in their cytoplasm, as an **energy source**.
5) The process of **forming glycogen** from glucose is called **glycogenesis**.

Liver cells are also called hepatocytes.

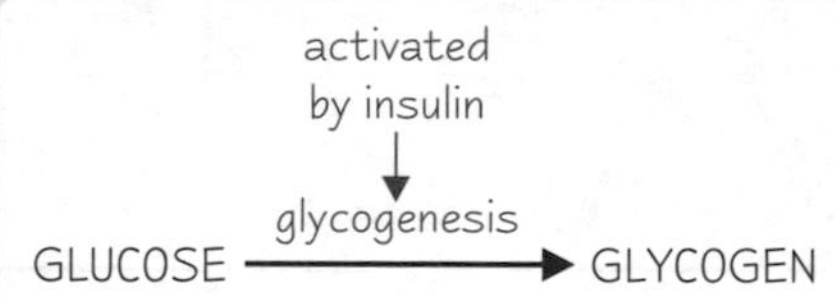

'Genesis' means 'making' — so glycogenesis means making glycogen.

6) Insulin also **increases** the **rate** of **respiration** of glucose, especially in muscle cells.

Glucagon raises blood glucose concentration when it's **too low**

1) Glucagon binds to **specific receptors** on the cell membranes of **liver cells**.
2) Glucagon **activates enzymes** that **break down glycogen** into **glucose**.
3) The process of **breaking down glycogen** is called **glycogenolysis**.
4) Glucagon also promotes the formation of glucose from **fatty acids** and **amino acids**.
5) The process of **forming glucose** from **non-carbohydrates** is called **gluconeogenesis**.

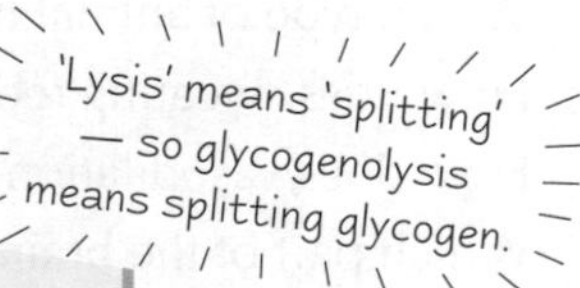

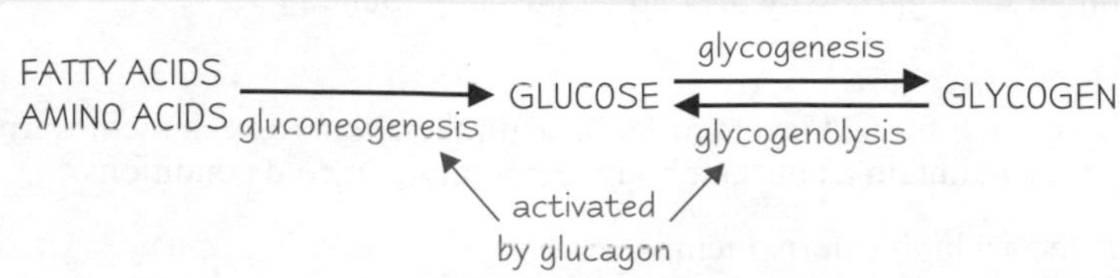

6) Glucagon **decreases** the **rate** of **respiration** of glucose in cells.

Melvin had finally mastered the ancient "chair-lysis" move.

Homeostasis — Control of Blood Glucose

Negative Feedback Mechanisms Keep Blood Glucose Concentration Normal

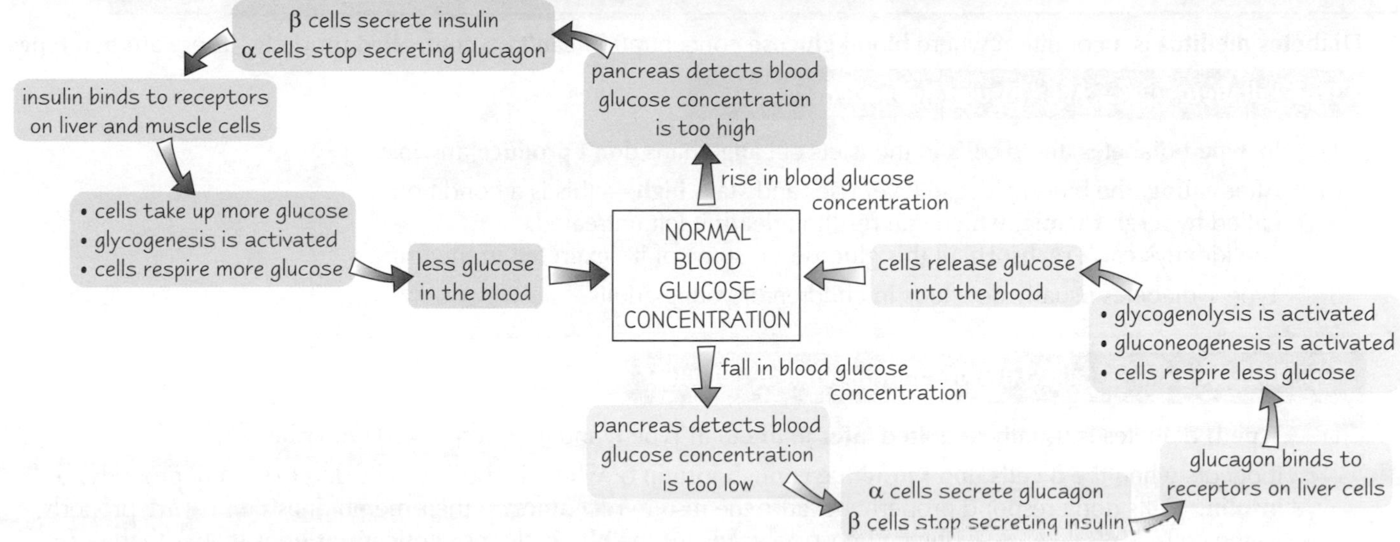

Beta (β) Cells Secrete Insulin when they're Depolarised

β cells **contain insulin** stored in **vesicles**. β cells **secrete insulin** when they **detect high blood glucose concentration**. Here's how it happens:

1) When blood glucose concentration is **high**, **more glucose enters** the β cells by **facilitated diffusion.**
2) **More glucose** in a β cell causes the rate of **respiration** to **increase**, making **more ATP**.
3) The **rise** in **ATP** triggers the **potassium ion channels** in the β cell plasma membrane to **close.**
4) This means **potassium ions** (K^+) **can't** get through the membrane — so they **build up inside** the cell.
5) This makes the **inside** of the β cell **less negative** because there are **more positively-charged** potassium ions **inside** the cell — so the plasma membrane of the β cell is **depolarised.**
6) Depolarisation triggers **calcium ion channels** in the membrane to **open**, so **calcium ions diffuse into** the β cell.
7) This causes the **vesicles** to **fuse** with the *β* **cell plasma membrane**, **releasing insulin** (this is called **exocytosis**).

Practice Questions

Q1 Why does your blood glucose concentration fall after exercise?

Q2 What's the process of breaking down glycogen into glucose called?

Q3 Give two effects of glucagon on liver cells.

Exam Questions

Q1 Describe and explain how hormones return blood glucose concentration to normal after a meal. [5 marks]

Q2 Suggest the effect on a β cell of respiration being inhibited. [2 marks]

My α cells detect low glucose — urgent tea and biscuit break needed...

Aaaaargh there are so many stupidly complex names to learn and they all look and sound exactly the same to me. You can't even get away with sneakily misspelling them all in your exam — like writing 'glycusogen' or 'glucogenesisolysis'. Nope, examiners have been around for centuries, so I'm afraid old tricks like that just won't work on them. Grrrrrrrr.

Diabetes and Control of Heart Rate

Phew, finally the last two pages in the section. Just a couple of loose ends to tidy up and you're home free.

Diabetes Occurs when Blood Glucose Concentration is Not Controlled

Diabetes mellitus is a condition where **blood glucose** concentration **can't** be **controlled** properly. There are **two types**:

Type I diabetes (also called **insulin-dependent** diabetes)

1) In **Type I** diabetes, the **β cells** in the islets of Langerhans **don't produce** any **insulin**.
2) After **eating**, the blood glucose level **rises** and **stays high** — this is a condition called **hyperglycaemia**, which can result in **death** if left untreated.
   The kidneys **can't reabsorb** all this glucose, so some of it's **excreted** in the urine.
3) Type 1 diabetes usually develops in children or young adults.

Type II diabetes (also called **non-insulin-dependent** diabetes)

1) **Type II** diabetes is usually acquired **later in life** than Type I, and it's often linked with **obesity**.
2) It occurs when the β cells **don't produce enough insulin** or when the body's **cells don't respond** properly to **insulin**. Cells don't respond properly because the insulin **receptors** on their membranes **don't work** properly, so the cells **don't** take up enough glucose. This means the **blood glucose concentration** is **higher** than normal.

Insulin can be Produced by Genetically Modified Bacteria

1) Insulin **used** to be **extracted** from **animal pancreases** (e.g. **pigs** and **cattle**), to treat people with **Type I** diabetes.
2) But **nowadays**, **human insulin** can be made by **genetically modified** (**GM**) **bacteria** (see p. 80).
3) Using **GM bacteria** to produce insulin is **much better** for many reasons, for example:
   - **Producing** insulin using GM bacteria is **cheaper** than extracting it from animal pancreases.
   - **Larger quantities** of insulin can be produced using GM bacteria.
   - GM bacteria make **human insulin**. This is **more effective** than using **pig** or **cattle insulin** (which is slightly different to human insulin) and it's **less likely** to trigger an **allergic response** or be **rejected** by the **immune system**.
   - Some people **prefer** insulin from **GM bacteria** for **ethical** or **religious** reasons. E.g. some **vegetarians** may **object** to the **use** of **animals**, and some **religious people object** to using insulin from **pigs**.

Stem Cells Could be Used to Cure Diabetes

1) Stem cells are **unspecialised cells** — they have the **ability** to **develop** into **any type** of cell.
2) Using stem cells could **potentially cure** diabetes — here's how:
   - **Stem cells** could be **grown** into **β cells**.
   - The β cells would then be **implanted** into the **pancreas** of a person with **Type I diabetes**.
   - This means the person would be able to **make insulin** as **normal**.
   - This treatment is **still being developed**. But if it's effective, it'll **cure** people with Type I diabetes — they **won't** have to have **regular injections** of **insulin**.

Look back at your AS notes if you need to remind yourself about stem cells.

Diabetes and Control of Heart Rate

The *Control* of *Heart Rate* Involves *Both* the *Nervous* and *Hormonal Systems*

The **nervous system** helps to control heart rate in these ways:

1) The **sinoatrial node** (**SAN**) generates **electrical impulses** that cause the **cardiac muscles** to **contract**.
2) The **rate** at which the SAN fires (i.e. heart rate) is **unconsciously controlled** by a part of the **brain** called the **medulla**.
3) Animals need to **alter** their **heart rate** to **respond** to **internal stimuli**, e.g. to prevent fainting due to low blood pressure or to make sure the heart rate is high enough to supply the body with enough oxygen.
4) **Stimuli** are **detected** by **pressure receptors** and **chemical receptors**:
   - There are **pressure receptors** called **baroreceptors** in the **aorta** and the **vena cava**. They're stimulated by **high** and **low blood pressure**.
   - There are **chemical receptors** called **chemoreceptors** in the **aorta**, the **carotid artery** (a major artery in the neck) and in the **medulla**. They **monitor** the **oxygen** level in the **blood** and also **carbon dioxide** and **pH** (which are indicators of O_2 level).
5) Electrical impulses from receptors are sent **to the medulla** along **sensory** neurones. The medulla processes the information and sends impulses to the SAN along **motor** neurones. Here's how it all works:

STIMULUS	RECEPTOR	NEURONE	EFFECTOR	RESPONSE
High blood pressure.	Baroreceptors detect high blood pressure.	Impulses are sent to the medulla, which sends impulses along the vagus nerve. This secretes acetylcholine, which binds to receptors on the SAN.	Cardiac muscles	Heart rate slows down to reduce blood pressure back to normal.
Low blood pressure.	Baroreceptors detect low blood pressure.	Impulses are sent to the medulla, which sends impulses along the accelerator nerve. This secretes noradrenaline, which binds to receptors on the SAN.	Cardiac muscles	Heart rate speeds up to increase blood pressure back to normal.
High blood O_2, low CO_2 or high pH levels.	Chemoreceptors detect chemical changes in the blood.	Impulses are sent to the medulla, which sends impulses along the vagus nerve. This secretes acetylcholine, which binds to receptors on the SAN.	Cardiac muscles	Heart rate decreases to return O_2, CO_2 and pH levels back to normal.
Low blood O_2, high CO_2 or low pH levels.	Chemoreceptors detect chemical changes in the blood.	Impulses are sent to the medulla, which sends impulses along the accelerator nerve. This secretes noradrenaline, which binds to receptors on the SAN.	Cardiac muscles	Heart rate increases to return O_2, CO_2 and pH levels back to normal.

The **hormonal system** helps to control heart rate in these ways:

1) When an organism is **threatened** (e.g. by a predator) the **adrenal glands** release **adrenaline**.
2) Adrenaline **binds** to **specific receptors** in the **heart**. This causes the cardiac muscle to **contract more frequently** and with **more force**, so **heart rate increases** and the heart **pumps more blood**.

Practice Questions

Q1 What is diabetes?

Q2 Briefly describe how stem cells could be used to cure diabetes.

Exam Questions

Q1 Explain why someone with diabetes can produce insulin but can't control their blood glucose concentration. [3 marks]

Q2 Give two advantages of using insulin produced by genetically modified (GM) bacteria over using insulin extracted from animal pancreases. [2 marks]

Q3 a) Explain how high blood pressure in the aorta causes the heart rate to slow down. [5 marks]
b) What would be the effect of severing the nerves from the medulla to the sinoatrial node (SAN)? [2 marks]

My heart rate seems to be controlled by the boy next door...

So, the hormonal system can work with the nervous system to control processes such as heart rate. However, sometimes the hormonal system goes wrong and causes problems, like diabetes. Luckily advances in medical technology (e.g. synthetic insulin and stem cells) have helped to treat these problems. Congratulations, you've made it to the end of the first section.

The Liver and Excretion

Liver — not just what your grandparents eat with onions. The liver has loads of functions, but the one you need to know about is its job in excretion. It's great at breaking things down like excess amino acids and other harmful substances.

Excretion is the Removal of Waste Products from the Body

All the **chemical reactions** that happen in your cells make up your **metabolism**. Metabolism produces **waste products** — substances that **aren't needed** by the cells, such as **carbon dioxide** and **nitrogenous** (nitrogen-containing) **waste**. Many of these products are **toxic**, so if they were allowed to **build up** in the body they would cause **damage**. This is where **excretion** comes in. Excretion is the **removal** of the **waste products of metabolism** from the body.

For example, **carbon dioxide** is a waste product of **respiration**. **Too much** in the blood is toxic, so it's removed from the body by the **lungs** (e.g. in mammals) or **gills** (e.g. in fish). The lungs and gills act as **excretory organs**.

The Liver is Involved in Excretion

One of the functions of the **liver** is to **break down** metabolic waste products and other substances that can be **harmful**, like **drugs** and **alcohol**. They're broken down into **less harmful** products that can then be **excreted**.

You need to learn all the different **veins**, **arteries** and **ducts** connected to the liver:

1) The **hepatic artery** supplies the liver with **oxygenated blood** from the heart, so the liver has a good supply of **oxygen** for **respiration**, providing plenty of **energy**.
2) The **hepatic vein** takes **deoxygenated blood** away from the liver.
3) The **hepatic portal vein** brings blood from the **duodenum** and **ileum** (parts of the small intestine), so it's rich in the products of **digestion**. This means any ingested harmful substances are **filtered out** and **broken down straight away**.
4) The **bile duct** takes **bile** (a substance produced by the liver to **emulsify fats**) to the **gall bladder** to be **stored**.

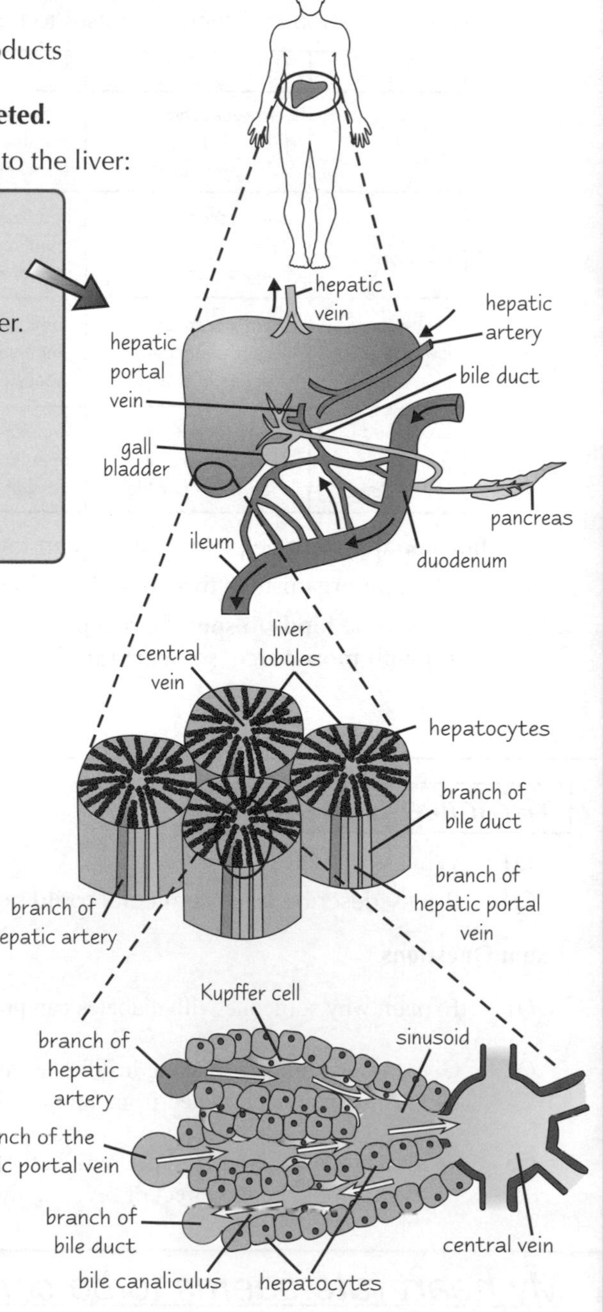

You need to learn about the **structure** of the liver too:

1) The liver is made up of **liver lobules** — **cylindrical** structures made of **cells** called **hepatocytes** that are arranged in rows **radiating** out from the centre.
2) Each lobule has a **central vein** in the middle that connects to the **hepatic vein**. **Many branches** of the **hepatic artery**, **hepatic portal vein** and **bile duct** are also found connected to each lobule (only one of each is shown in the picture).
3) The **hepatic artery** and the **hepatic portal vein** are connected to the **central vein** by **capillaries** called **sinusoids**.
4) Blood runs **through** the sinusoids, past the hepatocytes that **remove harmful substances** and **oxygen** from the blood.
5) The harmful substances are **broken down** by the hepatocytes into **less harmful** substances that then **re-enter** the blood.
6) The blood runs to the **central vein**, and the central veins from all the lobules **connect** up to form the **hepatic vein**.
7) Cells called **Kupffer cells** are also attached to the walls of the sinusoids. They **remove bacteria** and **break down** old **red blood cells**.
8) The **bile duct** is connected to the **central vein** by **tubes** called **canaliculi**.

The Liver and Excretion

Excess Amino Acids are Broken Down by the Liver

One of the liver's most important roles is getting rid of **excess amino acids** produced by eating and **digesting protein**. **Amino acids** contain **nitrogen** in their **amino groups**. **Nitrogenous substances can't** usually be **stored** by the body. This means **excess** amino acids can be **damaging** to the body, so they must be **used** by the body (e.g. to make proteins) or be **broken down and excreted**. Here's how excess amino acids are **broken down** in the **liver**:

1) First, the nitrogen-containing **amino groups** are **removed** from any **excess** amino acids, forming **ammonia** and **organic acids** — this process is called **deamination.**
2) The organic acids can be **respired** to give **ATP** or converted to **carbohydrate** and stored as **glycogen.**
3) Ammonia is **too toxic** for mammals to excrete directly, so it's **combined** with CO_2 in the **ornithine cycle** to create **urea.**
4) The urea is **released** from the liver into the **blood**. The **kidneys** then **filter** the blood and **remove** the urea as **urine** (see p. 22-23), which is excreted from the body.

Josie felt that warm feeling that meant a little bit of urea had just slipped out.

The Liver Removes Other Harmful Substances from the Blood

The **liver** also breaks down other harmful substances, like **alcohol**, **drugs** and **unwanted hormones**. They're broken down into **less harmful compounds** that can then be **excreted** from the body — this process is called **detoxification**. Some of the harmful products broken down by the liver include:

1) **Alcohol** (**ethanol**) — a **toxic** substance that can **damage** cells. It's **broken down** by the liver into **ethanal**, which is then broken down into a **less harmful** substance called **acetic acid**. **Excess** alcohol over a long period can lead to **cirrhosis** of the liver — this is when the cells of the liver **die** and **scar tissue blocks blood flow**.

2) **Paracetamol** — a common painkiller that's **broken down** by the liver. **Excess** paracetamol in the blood can lead to **liver** and **kidney failure**.

3) **Insulin** — a **hormone** that controls **blood glucose concentration** (see page 16). Insulin is also broken down by the liver as excess insulin can cause problems with blood sugar levels.

Practice Questions

Q1 Define excretion.

Q2 Why is excretion needed?

Q3 Which blood vessel brings oxygenated blood to the liver?

Q4 Name the blood vessel that brings blood to the liver from the small intestine.

Q5 What are liver lobules?

Exam Questions

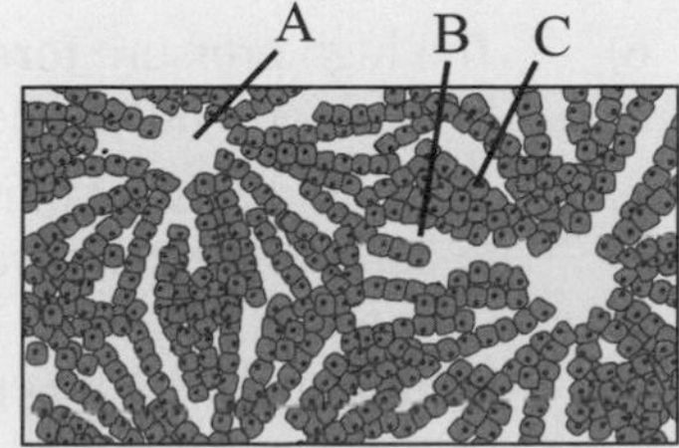

Q1 Name the parts of the liver shown in the diagram on the right. [3 marks]

Q2 Explain why the concentration of urea in urine might increase after eating a meal that's rich in protein. [6 marks]

Lots of important functions — can't liver without it...

Poor little amino acids, doing no harm then suddenly they're broken down and excreted. As upsetting as it is, however, you need to learn how they're broken down in the liver. It's a heart-wrenching tale of separation — the amino group and the organic acid are torn from each other's life. Right, enough of that nonsense. Learn it and learn it good.

The Kidneys and Excretion

So you've learnt about how the liver does a pretty good job at breaking down stuff for excretion. Now you get to learn that the kidneys like to play a part in this excretion malarkey too...

The Kidneys are Organs of Excretion

One of the main **functions** of the **kidneys** is to **excrete waste products**, e.g. **urea** produced by the **liver**. They also **regulate** the body's **water content** (see p. 24-25). Here's an overview of how they excrete waste products (you need to **learn** the **structure** of the kidneys too):

1) Blood **enters** the kidney through the **renal artery** and then passes through **capillaries** in the **cortex** of the kidneys.
2) As the blood passes through the capillaries, **substances** are **filtered out of the blood** and into **long tubules** that surround the capillaries. This process is called **ultrafiltration** (see below).
3) **Useful substances** (e.g. glucose) are **reabsorbed** back into the blood from the tubules in the **medulla** — this is called **selective reabsorption** (see next page).
4) The remaining **unwanted substances** (e.g. urea) pass along the tubules, then along the **ureter** to the **bladder**, where they're **expelled** as **urine**.
5) The filtered blood passes out of the kidneys through the **renal vein**.

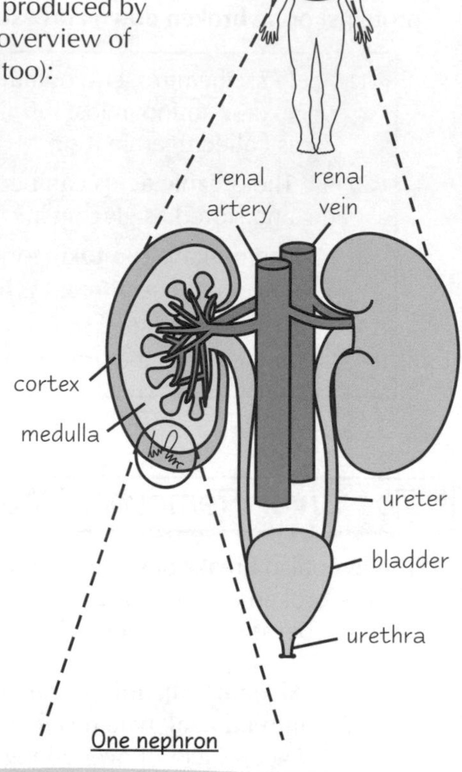

Blood is Filtered at the Start of the Nephrons

The **long tubules** along with the bundle of **capillaries** where the blood is **filtered** are called **nephrons** — there are **thousands** of nephrons in each kidney.

1) Blood from the **renal artery** enters smaller **arterioles** in the **cortex**.
2) Each arteriole splits into a structure called a **glomerulus** — a **bundle** of **capillaries** looped inside a hollow ball called a **renal capsule** (or **Bowman's capsule**).
3) This is where **ultrafiltration** takes place.
4) The **arteriole** that takes blood **into** each glomerulus is called the **afferent** arteriole, and the arteriole that takes the filtered blood **away** from the glomerulus is called the **efferent** arteriole.
5) The **efferent** arteriole is **smaller** in **diameter** than the afferent arteriole, so the blood in the glomerulus is under **high pressure**.
6) The high pressure **forces liquid** and **small molecules** in the blood **out** of the **capillary** and **into** the **renal capsule**.
7) The liquid and small molecules pass through **three** layers to get into the renal capsule and **enter** the nephron **tubules** — the **capillary wall**, a membrane (called the **basement membrane**) and the **epithelium** of the renal capsule. Larger molecules like **proteins** and **blood cells** **can't pass through** and **stay** in the blood.
8) The liquid and small molecules, now called **filtrate**, pass along the rest of the nephron and **useful substances** are **reabsorbed** along the way — see next page.
9) Finally, the filtrate flows through the **collecting duct** and passes out of the kidney along the **ureter**.

One nephron

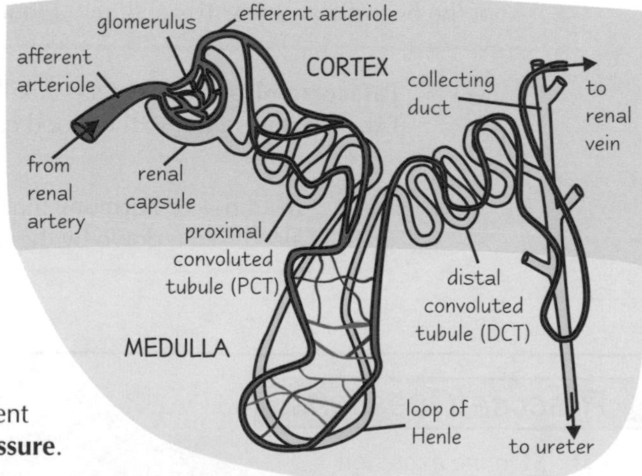

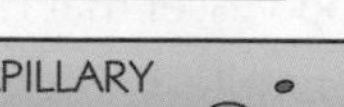

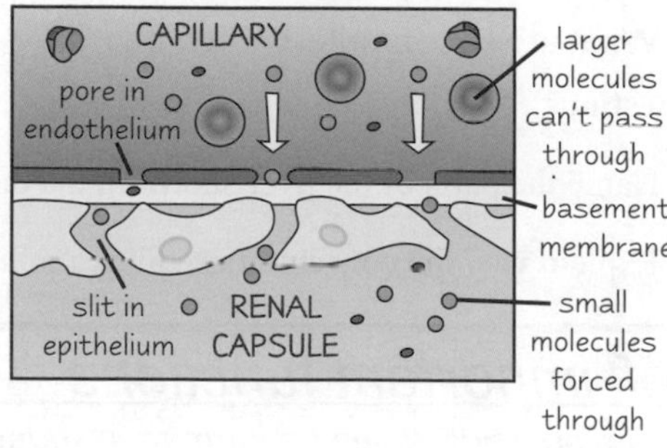

The Kidneys and Excretion

Useful Substances are **Reabsorbed** Along the **Nephron Tubules**

1) **Selective reabsorption** takes place as the filtrate flows along the **proximal convoluted tubule** (**PCT**), through the **loop of Henle**, and along the **distal convoluted tubule** (**DCT**).
2) **Useful substances** leave the tubules of the nephrons and **enter** the capillary network that's **wrapped** around them (see diagram on previous page).
3) The **epithelium** of the wall of the PCT has **microvilli** to provide a **large surface area** for the **reabsorption** of useful materials from the **filtrate** (in the tubules) into the **blood** (in the capillaries).
4) Useful solutes like **glucose**, **amino acids**, **vitamins** and some **salts** are reabsorbed along the PCT by **active transport** and **facilitated diffusion**.
5) Some **urea** is also reabsorbed by **diffusion**.
6) **Water** enters the blood by **osmosis** because the **water potential** of the blood is **lower** than that of the filtrate. Water is reabsorbed from the **loop of Henle**, **DCT** and the **collecting duct** (see next page).
7) The filtrate that remains is **urine**, which passes along the **ureter** to the **bladder**.

proximal convoluted tubule (PCT)
epithelial cell
microvilli
active transport
capillary

Water potential basically describes the tendency of water to move from one area to another. Water will move from an area of higher water potential to an area of lower water potential — it moves down the water potential gradient.

Urine is usually **made up of**:
- **Water** and **dissolved salts**.
- **Urea**.
- Other substances such as **hormones** and **excess vitamins**.

The volume of water in urine varies depending on how much you've drunk (see p. 24). The amount of urea also varies depending on how much protein you've eaten (see p. 21).

Urine **doesn't** usually contain:
- **Proteins** and **blood cells** — they're **too big** to be **filtered out** of the blood.
- **Glucose**, **amino acids** and **vitamins** — they're **actively reabsorbed** back into the blood (see above).

Ali was going to selectively absorb all the green jelly beans.

Practice Questions

Q1 Which blood vessel supplies the kidney with blood?

Q2 What are the bundles of capillaries found in the cortex of the kidneys called?

Q3 What is selective reabsorption?

Q4 Why aren't proteins normally found in urine?

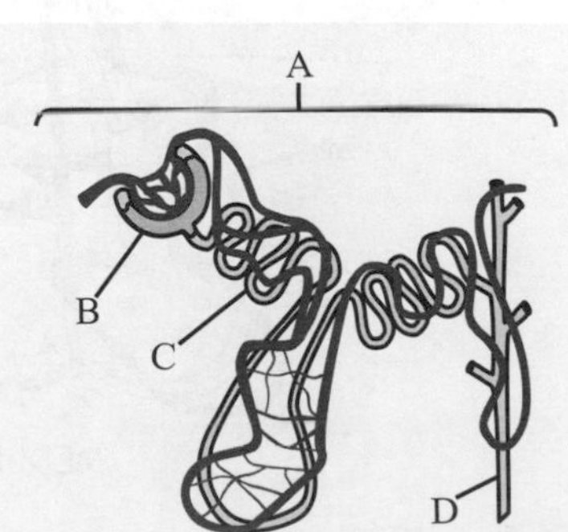

Exam Question

Q1 a) Name the structures labelled A-D shown in the diagram. [4 marks]

b) In which structure (B-D) does ultrafiltration take place? [1 mark]

c) Describe and explain the process of ultrafiltration. [5 marks]

Mmm — it's steak and excretion organ pie for dinner...

Excretion is a pretty horrible sounding word I know, but it's gotta be done. Speaking of horrible, I've never been able to eat kidney ever since I learnt all about this urine production business. Shame really because I used to love it sooooo much — I'd have kidneys on toast for breakfast, kidney sandwiches for lunch, kidney soup for tea, and kidney ice cream for pudding.

Controlling Water Content

More lovely kidney to gobble up on these pages — this time it's their role in controlling the water content of the blood. Busy things, these kidneys.

The **Kidneys** Regulate the **Water Content** of the **Blood**

Water is **essential** to keep the body **functioning**, so the **amount** of water in the **blood** needs to be kept **constant**. Mammals excrete **urea** (and other waste products) in **solution**, which means **water** is **lost** during excretion. Water is also lost in **sweat**. The kidneys **regulate** the water content of the blood (and urine), so the body has just the **right amount**:

If the water content of the blood is too **low** (the body is **dehydrated**), **more** water is **reabsorbed** by osmosis **into** the blood from the tubules of the nephrons (see p. 22-23 for more). This means the urine is **more concentrated**, so **less** water is **lost** during excretion.

If the water content of the blood is too **high** (the body is too **hydrated**), **less** water is **reabsorbed** by osmosis **into** the blood from the tubules of the nephrons. This means the urine is **more dilute**, so **more** water is **lost** during excretion (see next page).

Brad liked his urine to be dilute.

Regulation of the water content of the blood takes place in the **middle** and **last parts** of the nephron — the **loop of Henle**, the **distal convoluted tubule** (DCT) and the **collecting duct** (see below). The **volume** of water reabsorbed is controlled by **hormones** (see next page).

The **Loop of Henle** has a **Countercurrent Multiplier Mechanism**

The **loop of Henle** is made up of two **'limbs'** — the **descending** limb and the **ascending** limb. They help set up a mechanism called the **countercurrent multiplier mechanism**. It's this mechanism that helps to **reabsorb water** back into the blood. Here's how it **works**:

KEY
- Water moves out by osmosis
- Na^+ and Cl^- move by diffusion
- Na^+ and Cl^- move by active transport

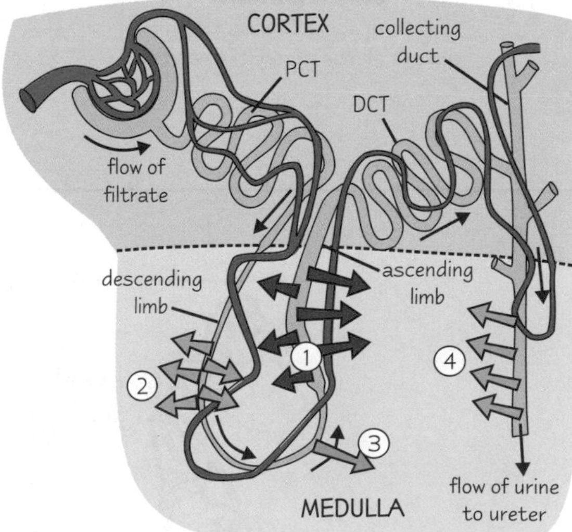

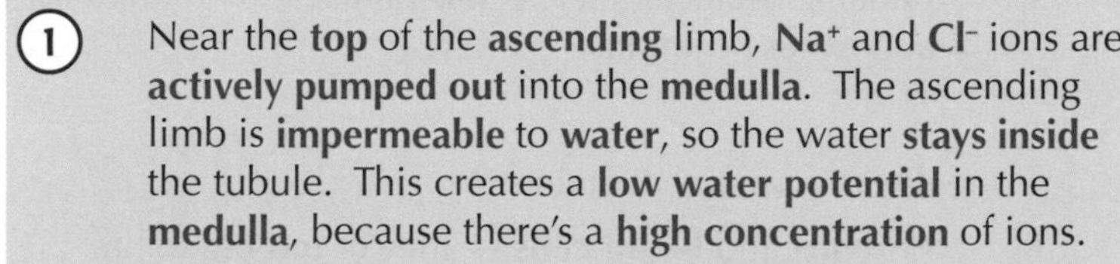

1. Near the **top** of the **ascending** limb, **Na^+** and **Cl^-** ions are **actively pumped out** into the **medulla**. The ascending limb is **impermeable** to **water**, so the water **stays inside** the tubule. This creates a **low water potential** in the **medulla**, because there's a **high concentration** of ions.

2. Because there's a **lower** water potential in the **medulla** than in the descending limb, **water** moves **out** of the **descending limb into** the **medulla** by **osmosis**. This makes the **filtrate more concentrated** (the ions can't diffuse out — the descending limb isn't permeable to them). The water in the medulla is **reabsorbed** into the **blood** through the **capillary network**.

3. Near the **bottom** of the **ascending** limb Na^+ and Cl^- ions **diffuse out** into the **medulla**, further **lowering** the **water potential** in the medulla. (The ascending limb is **impermeable** to **water**, so it **stays in the tubule**.)

4. The first three stages massively **increase** the **ion concentration** in the **medulla**, which **lowers** the **water potential**. This causes **water** to **move out** of the **collecting duct** by **osmosis**. As before, the water in the medulla is **reabsorbed** into the **blood** through the **capillary network**.

The **volume** of water **reabsorbed** from the collecting duct into the capillaries is **controlled** by **changing the permeability** of the **collecting duct** (see next page).

Different animals have **different length loops of Henle**. The **longer** an animal's loop of Henle, the **more water they can reabsorb** from the filtrate. When there's a longer ascending limb, **more ions** are **actively pumped out** into the medulla, which creates a **really low water potential** in the medulla. This means **more water** moves **out** of the nephron and collecting duct **into** the **capillaries**, giving very **concentrated urine**. Animals that live in areas where there's **little water** usually have **long loops** to **save** as much **water** as possible.

Controlling Water Content

Water Reabsorption is Controlled by Hormones

It's called antidiuretic hormone because diuresis is when lots of dilute urine is produced, so anti means a small amount of concentrated urine is produced.

1) The water content, and so water potential, of the blood is **monitored** by cells called **osmoreceptors** in a part of the **brain** called the **hypothalamus**.
2) When the osmoreceptors are **stimulated** by **low** water content in the blood, the hypothalamus sends **nerve impulses** to the **posterior pituitary gland** to release a **hormone** called **antidiuretic hormone** (ADH) into the blood.
3) ADH makes the walls of the DCT and collecting duct **more permeable** to **water**.
4) This means **more water** is **reabsorbed** from these tubules **into** the medulla and into the blood by osmosis. A **small** amount of **concentrated urine** is produced, which means **less water** is **lost** from the body.

Here's how ADH changes the **water content** of the **blood** when it's too **low** or too **high**:

1 Blood ADH Level Rises When You're Dehydrated

Dehydration is what happens when you **lose water**, e.g. by sweating during exercise, so the **water content** of the blood needs to be **increased**:

1) The **water content** of the blood **drops**, so its **water potential drops**.
2) This is detected by **osmoreceptors** in the **hypothalamus**.
3) The **posterior pituitary gland** is stimulated to release **more ADH** into the blood.
4) **More ADH** means that the DCT and collecting duct are **more permeable**, so **more water** is **reabsorbed** into the blood by osmosis.
5) A **small amount** of **highly concentrated** urine is produced and **less water** is **lost**.

Dehydrated? Me? As if...

2 Blood ADH Level Falls When You're Hydrated

If you're **hydrated**, you've taken in **lots of water**, so the **water content** of the blood needs to be **reduced**:

1) The **water content** of the blood **rises**, so its **water potential rises**.
2) This is detected by the **osmoreceptors** in the **hypothalamus**.
3) The **posterior pituitary gland** releases **less ADH** into the blood.
4) **Less ADH** means that the DCT and collecting duct are **less permeable**, so **less water** is **reabsorbed** into the blood by osmosis.
5) A **large amount** of **dilute** urine is produced and **more water** is **lost**.

Practice Questions

Q1 In which parts of the nephron does water reabsorption take place?

Q2 Describe what happens along the descending limb of the loop of Henle.

Q3 Which cells monitor the water content of the blood?

Q4 Which gland releases ADH?

Exam Questions

Q1 Describe and explain how water is reabsorbed into the capillaries from the nephron. [6 marks]

Q2 The level of ADH in the blood rises during strenuous exercise.
Explain the cause of the increase and the effect it has on kidney function. [6 marks]

If you don't understand what ADH does, ur-ine trouble...

Seriously though, there are two main things to learn from these pages — how water is reabsorbed from the tubules in the kidney, and how the water content of the blood is regulated by osmoreceptors, the hypothalamus and the posterior pituitary gland. Keep writing it down until you've got it sorted in your head, and you'll be just fine. Now I need a wee.

Kidney Failure and Detecting Hormones

Everything's fine while the kidneys are working well, but when they get damaged things don't run quite so smoothly.

Kidney Failure is When the Kidneys *Stop Working Properly*

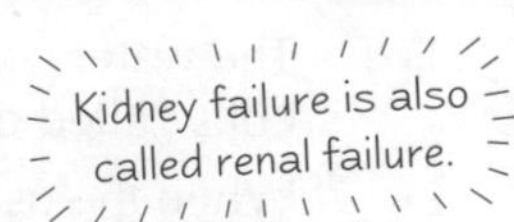

Kidney failure is when the kidneys **can't** carry out their **normal functions** because they **can't work properly**. Kidney failure can be **caused** by many things including:

1) **Kidney infections** — these can cause **inflammation** (swelling) of the kidneys, which can **damage** the cells. This **interferes** with **filtering** in the renal capsules, or with **reabsorption** in the other parts of the nephrons.
2) **High blood pressure** — this can damage the **glomeruli**. The blood in the glomeruli is already under **high pressure** but the **capillaries** can be **damaged** if the blood pressure gets **too high**. This means **larger** molecules like **proteins** can get through the capillary walls and into the **urine**.

Kidney failure causes **lots of problems**, for example:

1) **Waste products** that the kidneys would normally **remove** (e.g. **urea**) begin to **build up** in the blood. **Too much** urea in the blood causes **weight loss** and **vomiting**.
2) **Fluid** starts to **accumulate** in the tissues because the kidneys **can't remove excess water** from the blood. This causes **parts of the body** to **swell**, e.g. the person's legs, face and abdomen can swell up.
3) The balance of **ions** in the body becomes, well, unbalanced. The blood may become **too acidic**, and an imbalance of calcium and phosphate can lead to **brittle bones**. **Salt build-up** may cause more **water retention**.
4) **Long-term** kidney failure causes **anaemia** — a **lack** of **haemoglobin** in the blood.

If the problems caused by kidney failure **can't be controlled**, it can eventually lead to **death**.

Renal Dialysis and *Kidney Transplants* can be used to *Treat* Kidney Failure

When the kidneys can no longer **function** (i.e. they've **totally failed**), a person is unable to **survive** without **treatment**. There are **two** main treatment options:

Renal dialysis

1) **Renal dialysis** is where a **machine** is used to **filter** a patient's blood.
   - The patient's blood is passed through a **dialysis machine** — the **blood** flows on one side of a **partially permeable membrane** and **dialysis fluid** flows on the other side.
   - **Waste products** and **excess water** and **ions** diffuse across the membrane into the dialysis fluid, **removing** them from the blood.
   - **Blood cells** and **larger** molecules like **proteins** are **prevented** from **leaving** the blood.
2) Patients can feel increasingly **unwell** between dialysis sessions because **waste products** and **fluid** starts to build up in their **blood**.
3) Each dialysis session takes **three to five hours**, and patients need **two or three sessions a week**, usually in **hospital**. This is **quite expensive** and is pretty **inconvenient** for the patient.
4) But dialysis can keep a person **alive** until a **transplant** is available (see below), and it's a lot **less risky** than having the **major surgery** involved in a transplant.

Kidney transplant

1) A **kidney transplant** is where a **new kidney** is implanted into a patient's body to **replace** a damaged kidney.
2) The new kidney has to be from a person with the **same blood** and **tissue type**. They're often donated from a **living relative**, as people can survive with **only one** kidney. They can also come from **other people** who've recently **died** — organ donors.
3) Transplants have a lot of **advantages** over dialysis:
   - It's **cheaper** to give a person a transplant than keep them on dialysis for a **long time**.
   - It's **more convenient** for a person than regular dialysis sessions.
   - Patients don't have the problem of feeling **unwell** between dialysis sessions.
4) But there are also **disadvantages** to having a kidney transplant:
   - The patient will have to undergo a **major operation**, which is **risky**.
   - The **immune system** may **reject** the transplant, so the patient has to take **drugs** to **suppress it**.

Kidney Failure and Detecting Hormones

Urine is used to Test for Pregnancy and Steroid Use

Urine is made by **filtering** the **blood**, so you can have a look at what's in a person's blood by **testing** their **urine**. For example, you can test if a woman is **pregnant** by looking for a **hormone** that only pregnant women produce, and you can test **athletes** for the presence of **banned drugs** like **steroids**:

TESTING FOR PREGNANCY

Pregnancy tests detect the hormone **human chorionic gonadotropin** (**hCG**) that's only found in the **urine** of **pregnant women**:

1) A **stick** is used with an **application area** that contains **antibodies for hCG** bound to a **coloured bead** (**blue**).
2) When urine is applied to the application area any hCG will **bind** to the antibody on the beads.
3) The urine **moves** up to the **test strip**, **carrying** the **beads** with it.
4) The test strip has **antibodies to hCG** stuck in place (**immobilised**).
5) If there **is hCG present** the test strip turns **blue** because the **immobilised** antibody binds to any **hCG** attached to the **blue** beads, concentrating the **blue beads** in that area. If **no hCG** is present, the beads will **pass through** the test area **without** binding to anything, and so it **won't** go blue.

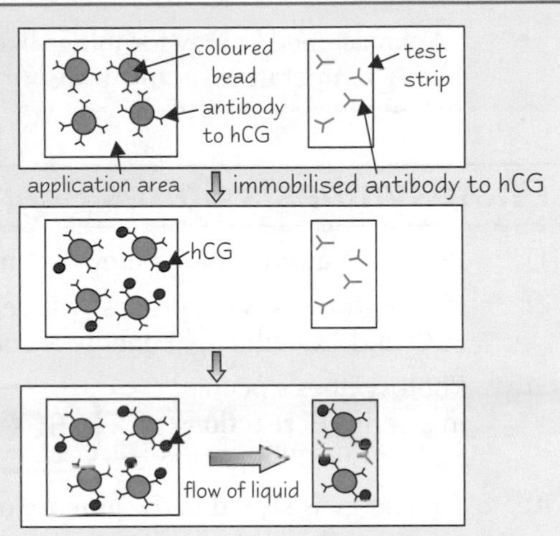

TESTING FOR STEROIDS

1) **Anabolic steroids** are **drugs** that **build up muscle tissue**.
2) **Testosterone** is an anabolic steroid, and there are other common ones such as **Nandrolone**.
3) Some **athletes** are **banned** from taking anabolic steroids. This is to try to stop the misuse of steroids that can have **dangerous side-effects**, such as **liver damage**. Also, it's considered **unfair** for some athletes to use steroids.
4) Steroids are **removed from the blood** in the **urine**, so athletes regularly have their **urine tested** for steroids.
5) Urine is tested for steroids (or the **products** made when they're **broken down**) by a technique called **gas chromatography**.
6) The urine sample is **vaporised** (turned into a **gas**) and passed through a column containing a **liquid**. **Different substances** move through the column at **different speeds**. The length of time taken for substances in the **sample** to pass through the column is **compared** to the time taken for a **steroid** to pass through the column. If the time taken is the **same** then the sample **contains the steroid**.

Practice Questions

Q1 Give one effect of kidney failure on the body.

Q2 What is renal dialysis?

Q3 What substance does a pregnancy test detect in a urine sample?

Exam Questions

Q1 Discuss the advantages and disadvantages of kidney transplants. [5 marks]

Q2 Describe and explain how urine can be used to detect steroid use. [5 marks]

Kidney failure, kidney infections, kidney transplants, kidney beans...

So you can either treat kidney failure with a kidney transplant or you can use kidney dialysis to filter the blood a few times a week. Both treatments come with their advantages and disadvantages, so make sure you can sum them both up. Here's a tip — you can usually use the disadvantages of one treatment to come up with the advantages of the other.

Photosynthesis, Respiration and ATP

OK, this isn't the easiest topic in the world, but 'cos I'm feeling nice today we'll take it slowly, one bit at a time...

Biological Processes Need Energy

Plant and animal cells **need energy** for biological processes to occur:

- **Plants** need energy for things like **photosynthesis**, **active transport** (e.g. to take in minerals via their roots), **DNA replication**, **cell division** and **protein synthesis**.
- **Animals** need energy for things like **muscle contraction**, maintenance of **body temperature**, **active transport**, **DNA replication**, **cell division** and **protein synthesis**.

Without energy, these biological processes would stop and the plant or animal would die.

Photosynthesis Stores Energy in Glucose

1) **Plants** are **autotrophs** — they can **make** their **own food** (**glucose**). They do this using **photosynthesis**.
2) **Photosynthesis** is the process where **energy** from **light** is used to **make glucose** from H_2O and CO_2 (the light energy is **converted** to **chemical energy** in the form of glucose).
3) Photosynthesis occurs in a **series** of **reactions**, but the overall equation is:

$$6CO_2 + 6H_2O + \text{Energy} \longrightarrow C_6H_{12}O_6 \text{ (glucose)} + 6O_2$$

4) So, energy is **stored** in the **glucose** until the plants **release** it by **respiration**.
5) **Animals** are **heterotrophs** — they **can't make** their **own food**. So, they obtain glucose by **eating plants** (or **other animals**), then respire the glucose to release energy.

Respiration in plants and animals needs glucose from photosynthesis to occur.

Cells Release Energy from Glucose by Respiration

1) **Plant** and **animal** cells **release energy** from **glucose** — this process is called **respiration**.
2) This energy is used to power all the **biological processes** in a cell.
3) There are two types of respiration:
   - **Aerobic respiration** — respiration **using oxygen**.
   - **Anaerobic respiration** — respiration **without oxygen**.
4) Aerobic respiration produces **carbon dioxide** and **water**, and releases **energy**. The overall equation is:

$$C_6H_{12}O_6 \text{ (glucose)} + 6O_2 \longrightarrow 6CO_2 + 6H_2O + \text{Energy}$$

ATP is the Immediate Source of Energy in a Cell

1) A cell **can't** get its energy **directly** from glucose.
2) So, in respiration, the **energy released** from glucose is used to **make ATP** (adenosine triphosphate). ATP is made from the nucleotide base **adenine**, combined with a **ribose sugar** and **three phosphate groups**.
3) It **carries energy** around the cell to where it's **needed**.
4) **ATP** is **synthesised** from **ADP** and **inorganic phosphate** (P_i) using energy from an **energy-releasing** reaction, e.g. the **breakdown** of **glucose** in **respiration**. The energy is stored as **chemical energy** in the **phosphate bond**. The enzyme **ATP synthase** catalyses this reaction.

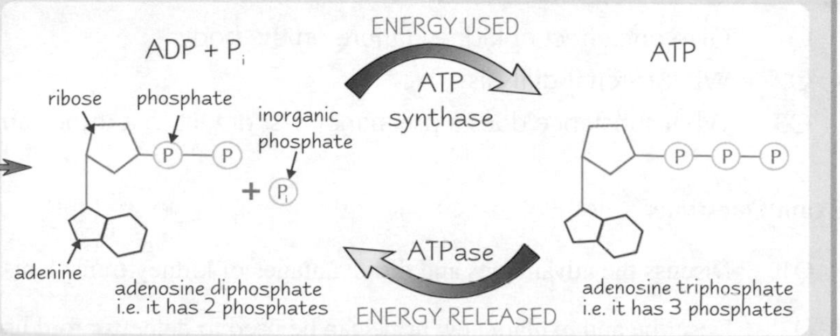

5) ATP **diffuses** to the part of the cell that **needs** energy.
6) Here, it's **broken down** back into **ADP** and **inorganic phosphate** (P_i). Chemical **energy** is **released** from the phosphate bond and used by the cell. **ATPase** catalyses this reaction.
7) The ADP and inorganic phosphate are **recycled** and the process starts again.

Inorganic phosphate (P_i) is just the fancy name for a single phosphate.

Photosynthesis, Respiration and ATP

*ATP has **Specific Properties** that Make it a **Good Energy Source***

1) ATP stores or releases only a **small, manageable amount** of energy at a time, so **no** energy is **wasted**.
2) It's a **small, soluble** molecule so it can be **easily transported** around the cell.
3) It's **easily broken down**, so energy can be **easily released**.
4) It can **transfer energy** to another molecule by transferring one of its **phosphate groups**.
5) ATP **can't pass out** of the **cell**, so the cell **always** has an immediate supply of energy.

Karen needed a lot of energy just to keep her headdress on...

*You Need to **Know Some Basics** Before You Start*

There are some pretty confusing technical terms in this section that you need to get your head around:

- **Metabolic pathway** — a **series** of **small reactions** controlled by **enzymes**, e.g. **respiration** and **photosynthesis**.
- **Phosphorylation** — **adding phosphate** to a molecule, e.g. **ADP** is phosphorylated to **ATP** (see previous page).
- **Photophosphorylation** — **adding phosphate** to a molecule using **light**.
- **Photolysis** — the **splitting** (lysis) of a molecule using **light** (photo) energy.
- **Hydrolysis** — the **splitting** (lysis) of a molecule using **water** (hydro).
- **Decarboxylation** — the **removal** of **carbon dioxide** from a molecule.
- **Dehydrogenation** — the **removal** of **hydrogen** from a molecule.
- **Redox reactions** — reactions that involve **oxidation** and **reduction**.

Remember redox reactions:

1) If something is **reduced** it has **gained electrons** (e^-), and may have **gained hydrogen** or lost oxygen.
2) If something is **oxidised** it has **lost electrons**, and may have **lost hydrogen** or gained oxygen.
3) Oxidation of one molecule **always** involves reduction of another molecule.

One way to remember electron and hydrogen movement is OILRIG. Oxidation Is Loss, Reduction Is Gain.

Photosynthesis** and **Respiration** Involve **Coenzymes

1) A **coenzyme** is a molecule that **aids** the **function** of an **enzyme**.
2) They work by **transferring** a **chemical group** from one molecule to another.
3) A coenzyme used in **photosynthesis** is **NADP**. NADP transfers **hydrogen** from one molecule to another — this means it can **reduce** (give hydrogen to) or **oxidise** (take hydrogen from) a molecule.
4) Examples of coenzymes used in **respiration** are: **NAD**, **coenzyme A** and **FAD**.
   - NAD and FAD transfer **hydrogen** from one molecule to another — this means they can **reduce** (give hydrogen to) or **oxidise** (take hydrogen from) a molecule.
   - **Coenzyme A** transfers **acetate** between molecules (see pages 39-40).

When hydrogen is transferred between molecules, electrons are transferred too.

Practice Questions

Q1 Write down three biological processes in animals that need energy.

Q2 What is photosynthesis?

Q3 What is the overall equation for aerobic respiration?

Q4 How many phosphate groups does ATP have?

Q5 Give the name of a coenzyme involved in photosynthesis.

Exam Question

Q1 ATP is the immediate source of energy inside a cell.
Describe how the synthesis and breakdown of ATP meets the energy needs of a cell. [6 marks]

Oh dear, I've used up all my ATP on these two pages...

Well, I won't beat about the bush, this stuff is pretty tricky... nearly as hard as a cross between Mr T, Hulk Hogan and Arnie. But, with a little patience and perseverance (and plenty of [chocolate] [coffee] [marshmallows] — delete as you wish), you'll get there. Once you've got these pages straight in your head, the next ones will be easier to understand.

Photosynthesis

Right, pen at the ready. Check. Brain switched on. Check. Cuppa piping hot. Check. Sweets on standby. Check. Okay, I think you're all sorted to start photosynthesis. Finally, take a deep breath and here we go...

Photosynthesis Takes Place in the Chloroplasts of Plant Cells

1) **Chloroplasts** are **small, flattened organelles** found in **plant cells**.
2) They have a **double membrane** called the **chloroplast envelope**.
3) **Thylakoids** (fluid-filled sacs) are **stacked up** in the chloroplast into structures called **grana** (singular = **granum**). The grana are **linked** together by bits of thylakoid membrane called **lamellae** (singular = **lamella**).
4) Chloroplasts contain **photosynthetic pigments** (e.g. **chlorophyll a, chlorophyll b** and **carotene**). These are **coloured substances** that **absorb** the **light energy** needed for photosynthesis. The pigments are found in the **thylakoid membranes** — they're attached to **proteins**. The protein and pigment is called a **photosystem**.
5) A photosystem contains **two types** of photosynthetic pigments — **primary** pigments and **accessory** pigments. Primary pigments are **reaction centres** where **electrons** are **excited** during the light-dependent reaction (see next page). Accessory pigments **surround** the primary pigments and **transfer light energy** to them.
6) There are **two** photosystems used by plants to capture light energy. **Photosystem I** (or PSI) absorbs light best at a wavelength of **700 nm** and **photosystem II** (PSII) absorbs light best at **680 nm**.

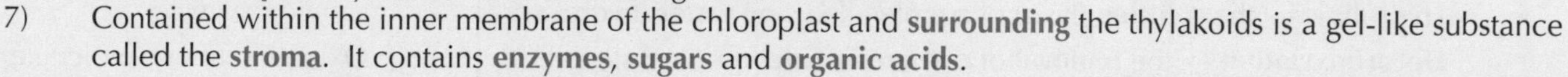

7) Contained within the inner membrane of the chloroplast and **surrounding** the thylakoids is a gel-like substance called the **stroma**. It contains **enzymes, sugars** and **organic acids**.
8) Carbohydrates produced by photosynthesis and not used straight away are stored as **starch grains** in the **stroma**.

Photosynthesis can be Split into Two Stages

There are actually **two stages** that make up **photosynthesis**:

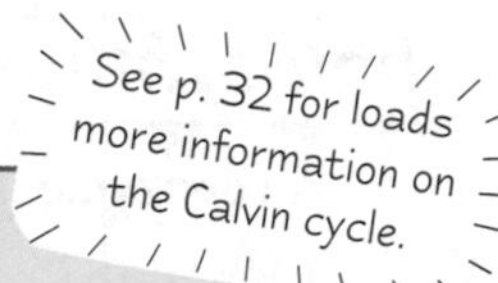

① The Light-Dependent Reaction

1) As the name suggests, this reaction **needs light energy**.
2) It takes place in the **thylakoid membranes** of the chloroplasts.
3) Here, light energy is absorbed by **photosynthetic pigments** in the **photosystems** and converted to **chemical energy**.
4) The light energy is used to add a phosphate group to ADP to form **ATP**, and to reduce NADP to form **reduced NADP**. **ATP transfers energy** and reduced **NADP transfers hydrogen** to the light-independent reaction.
5) During the process H_2O is **oxidised** to O_2.

② The Light-Independent Reaction

1) This is also called the **Calvin cycle** and as the name suggests it **doesn't use light energy** directly. (But it does **rely** on the **products** of the light-dependent reaction.)
2) It takes place in the **stroma** of the chloroplasts.
3) Here, the **ATP** and **reduced NADP** from the light-dependent reaction supply the **energy** and **hydrogen** to make **glucose** from CO_2.

In the Light-Dependent Reaction ATP is Made by Photophosphorylation

In the light-dependent reaction, the **light energy** absorbed by the photosystems is used for **three** things:

1) Making **ATP** from **ADP** and **inorganic phosphate**. This reaction is called **photophosphorylation** (see p. 29).
2) Making **reduced NADP** from **NADP**.
3) Splitting **water** into **protons** (H^+ ions), **electrons** and **oxygen**. This is called **photolysis** (see p. 29).

The light-dependent reaction actually includes **two types** of **photophosphorylation** — **non-cyclic** and **cyclic**. Each of these processes has **different products**.

Photosynthesis

Non-cyclic Photophosphorylation Produces ATP, Reduced NADP and O_2

To understand the process you need to know that the photosystems (in the thylakoid membranes) are **linked** by **electron carriers**. Electron carriers are **proteins** that **transfer electrons**. The photosystems and electron carriers form an **electron transport chain** — a **chain** of **proteins** through which **excited electrons flow**. All the processes in the diagrams are happening together — I've just split them up to make it easier to understand.

1) Light energy excites electrons in chlorophyll

- **Light energy** is absorbed by **PSII**.
- The light energy **excites electrons** in **chlorophyll**.
- The electrons move to a **higher energy level** (i.e. they have more energy).
- These high-energy electrons **move along** the **electron transport chain** to **PSI**.

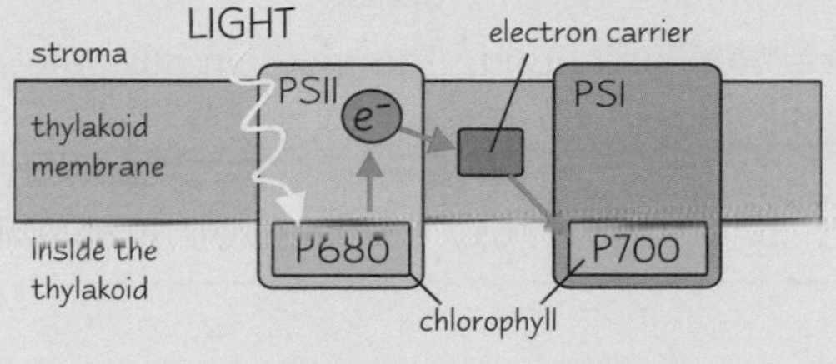

2) Photolysis of water produces protons (H^+ ions), electrons and O_2

- As the excited electrons **from chlorophyll leave PSII** to **move along** the electron transport chain, they must be **replaced**.
- **Light** energy splits **water** into **protons** (H^+ ions), **electrons** and **oxygen**. (So the O_2 in photosynthesis comes from water.)
- The reaction is: $H_2O \longrightarrow 2H^+ + \frac{1}{2}O_2$

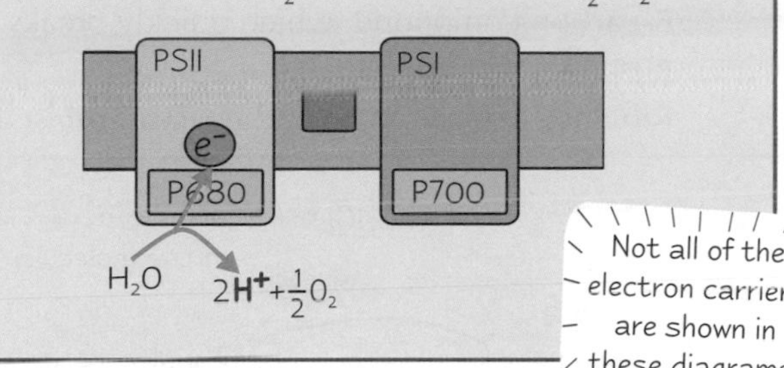

Not all of the electron carriers are shown in these diagrams.

3) Energy from the excited electrons makes ATP...

- The excited electrons **lose energy** as they **move along** the **electron transport chain**.
- This energy is used to **transport protons into** the **thylakoid** so that the thylakoid has a **higher concentration** of protons than the stroma. This forms a **proton gradient** across the membrane.
- Protons move **down** their concentration gradient, into the stroma, **via** an enzyme called **ATP synthase**. The energy from this movement combines **ADP** and **inorganic phosphate** (P_i) to form **ATP**.

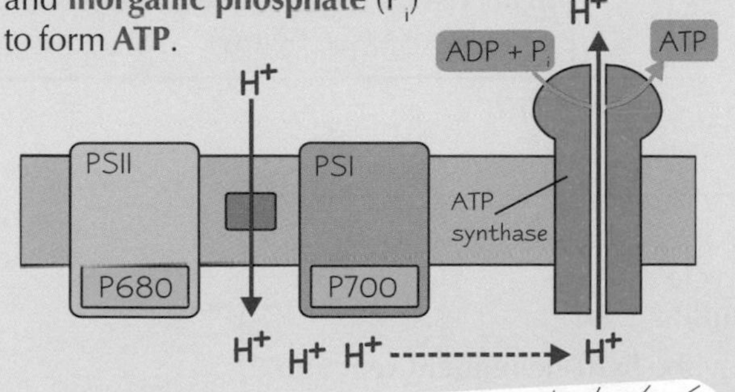

Chemiosmosis is the name of the process where the movement of H^+ ions across a membrane generates ATP. This process also occurs in respiration (see p. 41).

4) ...and generates reduced NADP.

- Light energy is **absorbed** by PSI, which excites the electrons again to an **even higher** energy level.
- Finally, the electrons are **transferred** to **NADP**, along with a **proton** (H^+ ion) from the **stroma**, to form **reduced NADP**.

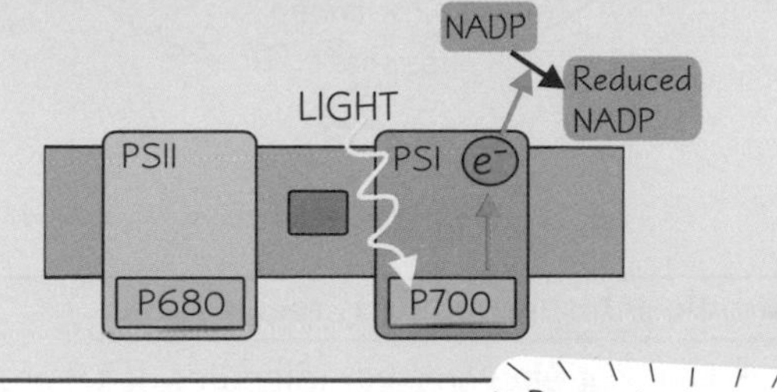

Remember a 'proton' is just another word for a hydrogen ion (H^+).

Cyclic Photophosphorylation Only Produces ATP

Cyclic photophosphorylation **only uses PSI**. It's called 'cyclic' because the electrons from the chlorophyll molecule **aren't** passed onto NADP, but are **passed back** to PSI via electron carriers. This means the electrons are **recycled** and can repeatedly flow through PSI. This process doesn't produce any reduced NADP or O_2 — it **only produces** small amounts of **ATP**.

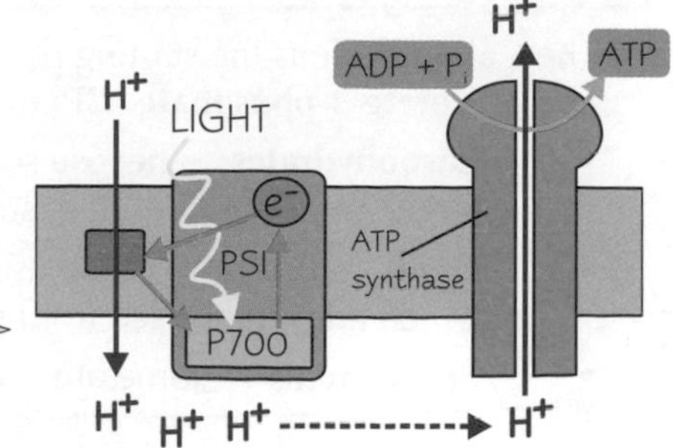

Photosynthesis

Don't worry, you're over the worst of photosynthesis now. Instead of electrons flying around, there's a nice cycle of reactions to learn. What more could you want from life? Money, fast cars and nice clothes have nothing on this...

The Light-Independent Reaction is also called the Calvin Cycle

1) The Calvin cycle takes place in the **stroma** of the chloroplasts.
2) It makes a molecule called **triose phosphate** from **CO_2** and **ribulose bisphosphate** (a 5-carbon compound). Triose phosphate can be used to make **glucose** and other **useful organic substances** (see below).
3) There are a few steps in the cycle, and it needs **ATP** and **H^+ ions** to keep it going.
4) The reactions are linked in a **cycle**, which means the starting compound, **ribulose bisphosphate**, is **regenerated**.

The Calvin cycle is also called carbon fixation, because carbon from CO_2 is 'fixed' into an organic molecule.

Here's what happens at each stage in the cycle:

1) Carbon dioxide is combined with ribulose bisphosphate to form two molecules of glycerate 3-phosphate

- CO_2 enters the leaf through the **stomata** and diffuses into the **stroma** of the chloroplast.
- Here, it's combined with **ribulose bisphosphate (RuBP)**, a **5-carbon** compound. This gives an **unstable 6-carbon** compound, which quickly breaks down into **two** molecules of a **3-carbon** compound called **glycerate 3-phosphate (GP)**.
- **Ribulose bisphosphate carboxylase (rubisco)** catalyses the reaction between CO_2 and **ribulose bisphosphate**.

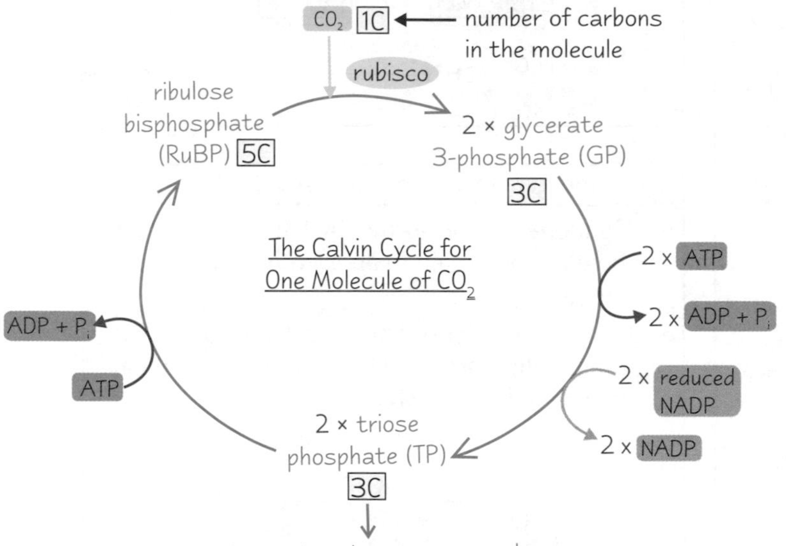

2) ATP and reduced NADP are required for the reduction of GP to triose phosphate

- Now **ATP**, from the **light-dependent** reaction, **provides energy** to turn the **3-carbon** compound, **GP**, into a **different** 3-carbon compound called **triose phosphate (TP)**.
- This reaction also requires **H^+ ions**, which come from **reduced NADP** (also from the **light-dependent reaction**). Reduced NADP is **recycled** to **NADP**.
- **Triose phosphate** is then converted into many **useful organic compounds**, e.g. glucose (see below).

3) Ribulose bisphosphate is regenerated

- **Five** out of every **six** molecules of **TP** produced in the cycle aren't used to make hexose sugars, but to **regenerate RuBP**.
- Regenerating RuBP uses the **rest** of the **ATP** produced by the **light-dependent reaction**.

TP and GP are Converted into Useful Organic Substances like Glucose

The Calvin cycle is the starting point for making **all** the organic substances a plant needs. **Triose phosphate** (TP) and **glycerate 3-phosphate** (GP) molecules are used to make **carbohydrates**, **lipids** and **amino acids**:

- **Carbohydrates** — **hexose sugars** (e.g. glucose) are made by joining **two triose phosphate molecules** together and **larger** carbohydrates (e.g. sucrose, starch, cellulose) are made by joining **hexose sugars** together in **different ways**.
- **Lipids** — these are made using **glycerol**, which is synthesised from **triose phosphate**, and **fatty acids**, which are synthesised from **glycerate 3-phosphate**.
- **Amino acids** — some amino acids are made from **glycerate 3-phosphate**.

Photosynthesis

The Calvin Cycle Needs to Turn Six Times to Make One Hexose Sugar

1) **Three turns** of the cycle produces **six** molecules of **triose phosphate** (TP), because two molecules of TP are made for every one CO_2 molecule used.
2) **Five** out of **six** of these TP molecules are used to **regenerate ribulose bisphosphate** (RuBP).
3) This means that for **three turns** of the cycle only **one TP** is produced that's used to make a **hexose sugar**.
4) A hexose sugar has **six carbons** though, so **two TP** molecules are needed to form one hexose sugar.
5) This means the cycle must turn **six times** to produce **two molecules** of **TP** that can be used to make **one hexose sugar**.
6) Six turns of the cycle need **18 ATP** and **12 reduced NADP** from the light-dependent reaction.

The Structure of a Chloroplast is Adapted for Photosynthesis

1) The **chloroplast envelope** keeps the **reactants** for photosynthesis **close** to their **reaction sites**.
2) The **thylakoids** have a **large surface area** to allow as much **light energy** to be **absorbed** as possible.
3) **Lots of ATP synthase** molecules are present in the thylakoid membranes to **produce ATP** in the light-dependent reaction.
4) The **stroma** contains all the **enzymes**, **sugars** and **organic acids** for the light-independent reaction to take place.

Practice Questions

Q1 Name two photosynthetic pigments in the chloroplasts of plants.

Q2 At what wavelength does photosystem I absorb light best?

Q3 What three substances does non-cyclic photophosphorylation produce?

Q4 Which photosystem is involved in cyclic photophosphorylation?

Q5 Where in the chloroplasts does the light-independent reaction occur?

Q6 Name two organic substances made from triose phosphate.

Q7 How many CO_2 molecules need to enter the Calvin cycle to make one hexose sugar?

Q8 Describe two ways in which a chloroplast is adapted for photosynthesis.

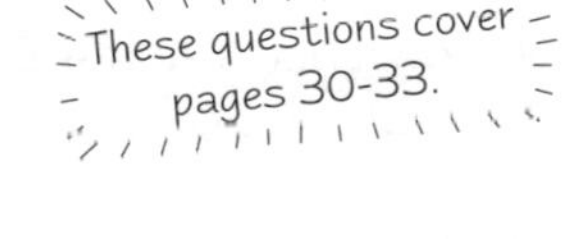

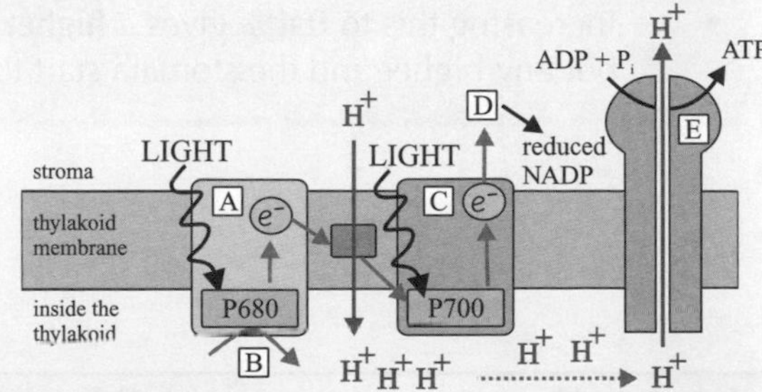

Exam Questions

Q1 The diagram above shows the light-dependent reaction of photosynthesis.
a) Where precisely in a plant does the light-dependent reaction of photosynthesis occur? [1 mark]
b) What is A? [1 mark]
c) Describe process B and explain its purpose. [4 marks]
d) What is reactant D? [1 mark]

Q2 Rubisco is an enzyme that catalyses the first reaction of the Calvin cycle. CA1P is an inhibitor of rubisco.
a) Describe how triose phosphate is produced in the Calvin cycle. [6 marks]
b) Briefly explain how ribulose bisphosphate (RuBP) is regenerated in the Calvin cycle. [2 marks]
c) Explain the effect that CA1P would have on glucose production. [3 marks]

Calvin cycles — bikes made by people that normally make pants...

Next thing we know there'll be male models swanning about in their pants riding highly fashionable bikes. Sounds awful I know, but let's face it, anything would look better than cycling shorts. Anyway, it would be a good idea to go over these pages a couple of times — you might not feel as if you can fit any more information in your head, but you can, I promise.

Limiting Factors in Photosynthesis

Now you know what photosynthesis is it's time to find out what conditions make it speedy and what slows it down. I'd start by making sure you have the best conditions for revision — oodles of biscuits and your thinking cap on.

There are **Optimum Conditions** for Photosynthesis

The **ideal conditions** for photosynthesis vary from one plant species to another, but the conditions below would be ideal for **most** plant species in temperate climates like the UK.

1. High light intensity of a certain **wavelength**

- Light is needed to provide the **energy** for the **light-dependent reaction** — the **higher** the **intensity** of the light, the **more energy** it provides.
- Only certain **wavelengths** of light are used for photosynthesis. The photosynthetic pigments chlorophyll a, chlorophyll b and carotene only **absorb** the **red** and **blue** light in sunlight. (**Green** light is **reflected**, which is why plants look green.)

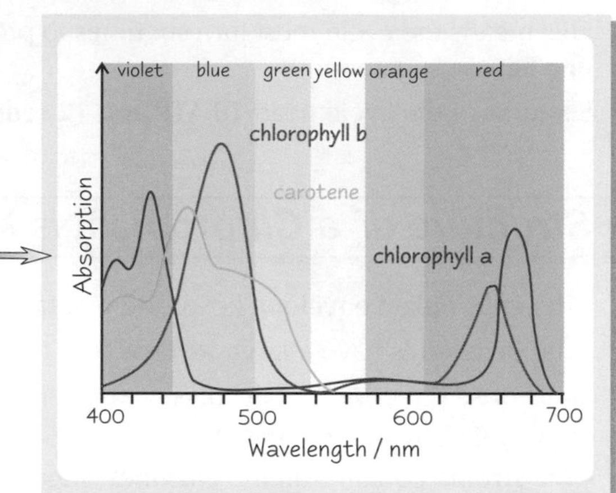

2. Temperature around **25 °C**

- Photosynthesis involves **enzymes** (e.g. ATP synthase, rubisco). If the temperature falls **below 10 °C** the enzymes become **inactive**, but if the temperature is **more than 45 °C** they may start to **denature**.
- Also, at **high** temperatures **stomata close** to avoid losing too much water. This causes photosynthesis to slow down because **less CO_2** enters the leaf when the stomata are closed.

3. Carbon dioxide at 0.4%

- Carbon dioxide makes up **0.04%** of the gases in the atmosphere.
- Increasing this to **0.4%** gives a **higher rate** of photosynthesis, but any higher and the stomata start to **close**.

Plants also need a constant supply of water — too little and photosynthesis has to stop but too much and the soil becomes waterlogged (reducing the uptake of magnesium for chlorophyll a).

Light, **Temperature** and CO_2 can all **Limit Photosynthesis**

1) **All three** of these things need to be at the **right level** to allow a plant to photosynthesise as quickly as possible.
2) If any **one** of these factors is **too low** or **too high**, it will **limit photosynthesis** (slow it down). Even if the other two factors are at the perfect level, it won't make **any difference** to the speed of photosynthesis as long as that factor is at the wrong level.
3) On a warm, sunny, windless day, it's usually CO_2 that's the limiting factor, and at night it's the **light intensity**.
4) However, **any** of these factors could become the limiting factor, depending on the **environmental conditions**.

All that Murray and Fraser knew was that limiting photosynthesis was a tasty business...

Limiting Factors in Photosynthesis

You Might Have to *Interpret Graphs* About *Limiting Factors*

Light intensity

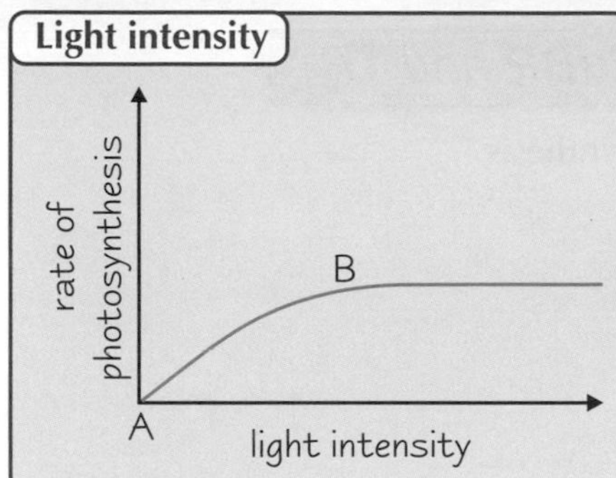

Between points A and B, the rate of photosynthesis is limited by the **light intensity**. So as the light intensity **increases**, so can the rate of photosynthesis. Point B is the **saturation point** — increasing light intensity after this point makes no difference, because **something else** has become the limiting factor. The graph now **levels off**.

The saturation point is where a factor is no longer limiting the reaction — something else has become the limiting factor.

Temperature

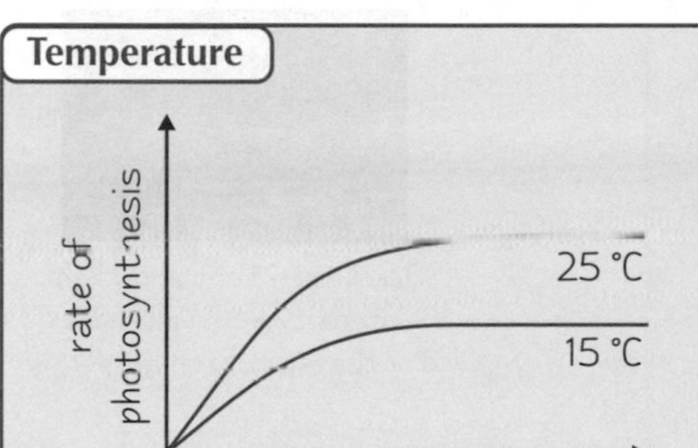

Both these graphs level off when **light intensity** is no longer the limiting factor. The graph at **25 °C** levels off at a **higher point** than the one at **15 °C**, showing that **temperature** must have been a limiting factor at **15 °C**.

CO_2 concentration

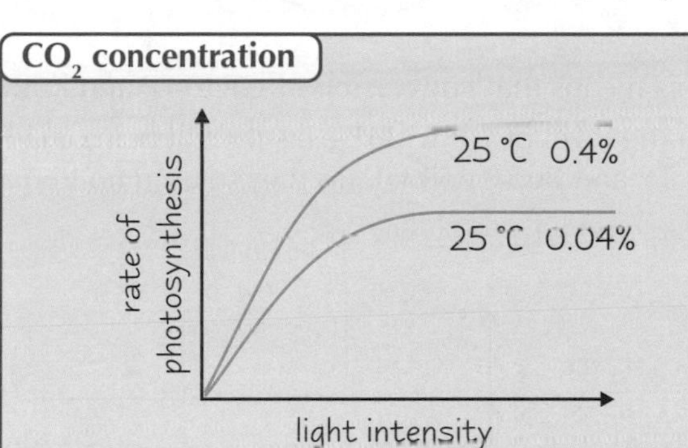

Again, both these graphs level off when **light intensity** is no longer the limiting factor. The graph at **0.4% CO_2** levels off at a **higher point** than the one at **0.04%**, so **CO_2 concentration** must have been a limiting factor at **0.04% CO_2**. The limiting factor here **isn't temperature** because it's the **same** for both graphs (25 °C).

Practice Questions

Q1 Why is a high light intensity an optimum condition for photosynthesis?

Q2 What is the optimum concentration of carbon dioxide for photosynthesis?

Q3 What is the limiting factor for photosynthesis on a warm, sunny day?

Q4 What is the limiting factor for photosynthesis at night?

Exam Question

Q1 An experiment was carried out to investigate how temperature affects photosynthesis.
The rate of photosynthesis was measured at 10 °C, 25 °C and 45 °C.
At which temperature would the rate of photosynthesis have been greatest? Explain your answer. [4 marks]

I'm a whizz at the factors that limit revision...

... watching Hollyoaks, making tea, watching EastEnders, walking the dog... not to mention staring into space (one of my favourites). These pages aren't that bad though. You just need to learn how light, CO_2 and temperature affect the rate of photosynthesis. Try shutting the book and writing down what you know — you'll be amazed at what you remember.

Limiting Factors in Photosynthesis

Well, I hope you didn't think we'd finished covering limiting factors.... ohhhhhh no, I could write a whole book on them. But just for you I've added an experiment to spice things up a bit. Woo hoo.

Light**, **Temperature** and **CO_2** Affect the Levels of **GP**, **RuBP** and **TP

Light intensity, **temperature** and **CO_2 concentration** all **affect** the **rate** of **photosynthesis**, which means they affect the **levels** of **GP**, **RuBP** and **TP** in the **Calvin cycle**.

1. Light intensity

- In **low light intensities**, the products of the light-dependent stage (**reduced NADP** and **ATP**) will be in **short supply**.

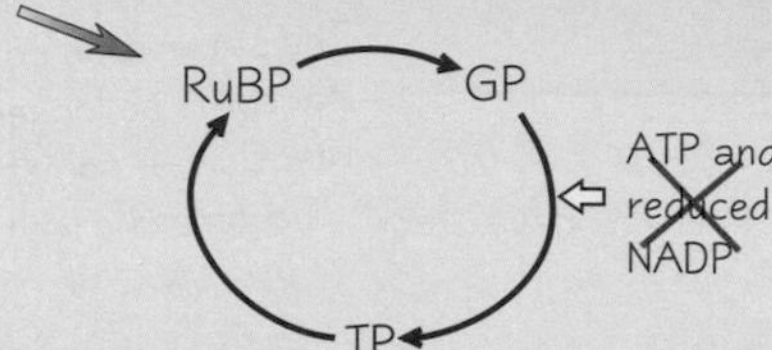

- This means that **conversion of GP** to TP and RuBP is **slow**.
- So the level of **GP** will **rise** (as it's still being made) and levels of **TP** and **RuBP** will **fall** (as they're used up to make GP).

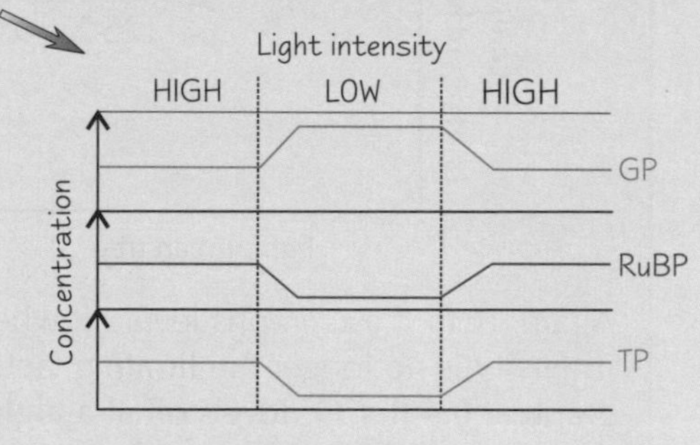

Derek knew that a low light intensity would increase the level of romance.

2. Temperature

- All the reactions in the Calvin cycle are catalysed by **enzymes** (e.g. rubisco).
- At **low temperatures**, all of the reactions will be **slower** as the enzymes work more **slowly**.

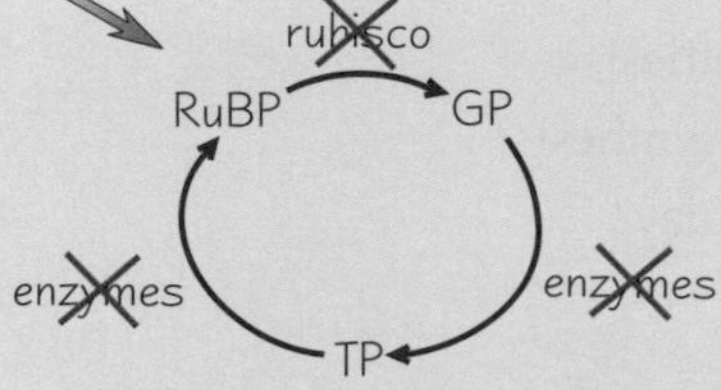

- This means the levels of **RuBP**, **GP** and **TP** will **fall**.
- GP, TP and RuBP are affected in the same way at **very high temperatures**, because the **enzymes** will start to **denature**.

3. Carbon dioxide concentration

- At **low CO_2 concentrations**, **conversion of RuBP** to GP is also **slow** (as there's less CO_2 to combine with RuBP to make GP).

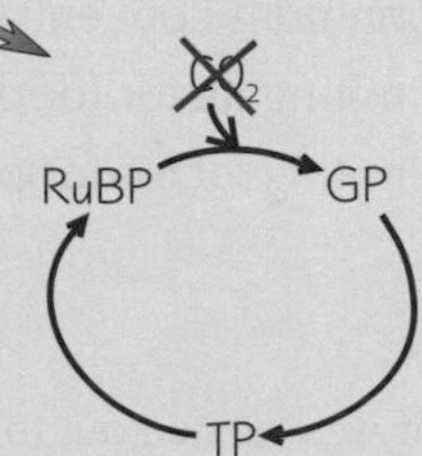

- So the level of **RuBP** will **rise** (as it's still being made) and levels of **GP** and **TP** will **fall** (as they're used up to make RuBP).

Limiting Factors in Photosynthesis

Limiting Factors can be Investigated using Pondweed

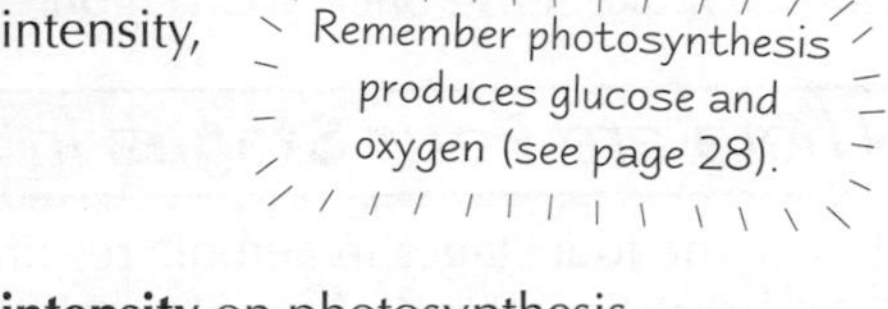

1) **Canadian pondweed** (*Elodea*) can be used to measure the effect of light intensity, temperature and CO_2 concentration on the **rate of photosynthesis**.
2) The rate at which **oxygen** is **produced** by the pondweed can be easily **measured** and this **corresponds** to the rate of photosynthesis.
3) For example, the **apparatus** below is used to **measure** the **effect** of **light intensity** on photosynthesis.

- A **test tube** containing the **pondweed** and **water** is connected to a **capillary tube** full of water.
- The tube of water is connected to a **syringe**.
- A **source of white light** is placed at a **specific distance** from the pondweed.
- The pondweed is left to photosynthesise for a **set** amount of **time**. As it photosynthesises, the **oxygen released** will **collect** in the **capillary tube**.
- At the end of the experiment, the syringe is used to **draw** the gas **bubble** in the tube **up** alongside a **ruler** and the **length** of the gas bubble (volume of O_2) is **measured**.
- Any **variables** that could affect the results should be **controlled**, e.g. temperature, the time the weed is left to photosynthesise.
- The experiment is **repeated** and the **average** length of gas bubble is calculated, to make the results **more reliable**.
- The whole experiment is then **repeated** with the **light source** placed at **different distances** from the pondweed.

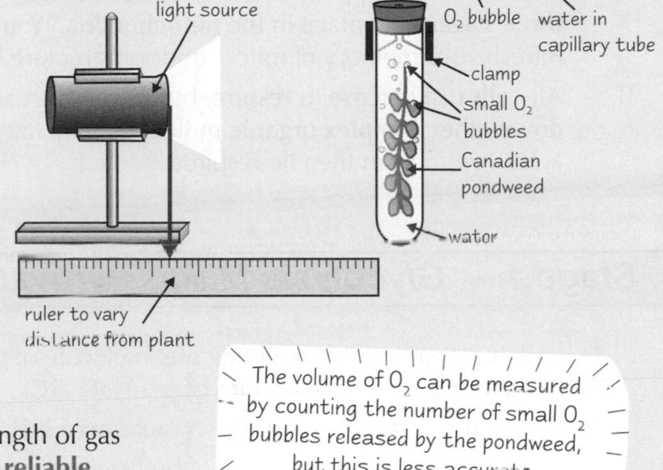

The volume of O_2 can be measured by counting the number of small O_2 bubbles released by the pondweed, but this is less accurate.

4) The apparatus above can be adapted to **measure** the **effect** of **temperature** on photosynthesis — the test tube of pondweed is put in a **beaker of water** at a **set temperature** (then the experiment's repeated with different temperatures of water).

Practice Questions

Q1 How does a low CO_2 concentration in the air affect the level of TP in a plant?

Q2 In an experiment on the rate of photosynthesis, how can light intensity be varied?

Q3 In the experiment above, give two variables that must be controlled.

Exam Questions

Q1 A scientist was investigating the effect of different conditions on the levels of GP, TP and RuBP in a plant. Predict the results of his experiment under the following conditions. Explain your answers.

a) Low light intensity, optimum temperature and optimum CO_2 concentration. [3 marks]

b) Low temperature, optimum light intensity and optimum CO_2 concentration. [3 marks]

Q2 Briefly describe the apparatus and method you would use to investigate how temperature affects photosynthesis in Canadian pondweed. [6 marks]

Aah, Canadian pondweed — a biology student's best friend...

Well... sometimes — usually you end up staring endlessly at it while it produces lots of tiny bubbles. Thrilling. If you have to describe an experiment in the exam make sure you include details about the apparatus, the method and the variables you'd keep constant to make your results more reliable. Examiners love reliability (but then they're a bit weird — I love cake).

Aerobic Respiration

From the last gazillion pages you know that plants make their own glucose. Unfortunately, that means now you need to learn how plant and animal cells release energy from glucose. It's not the easiest thing in the world to understand, but it'll make sense once you've gone through it a couple of times.

There are **Four Stages** in **Aerobic Respiration**

1) The four stages in aerobic respiration are **glycolysis**, the **link reaction**, the **Krebs cycle** and **oxidative phosphorylation**.
2) The **first three** stages are a **series of reactions**. The **products** from these reactions are **used** in the **final stage** to produce loads of ATP.
3) The **first** stage happens in the **cytoplasm** of cells and the **other three** stages take place in the **mitochondria**. You might want to refresh your memory of mitochondrion structure before you start.
4) All cells use **glucose** to **respire**, but organisms can also **break down** other **complex organic molecules** (e.g. fatty acids, amino acids), which can then be respired.

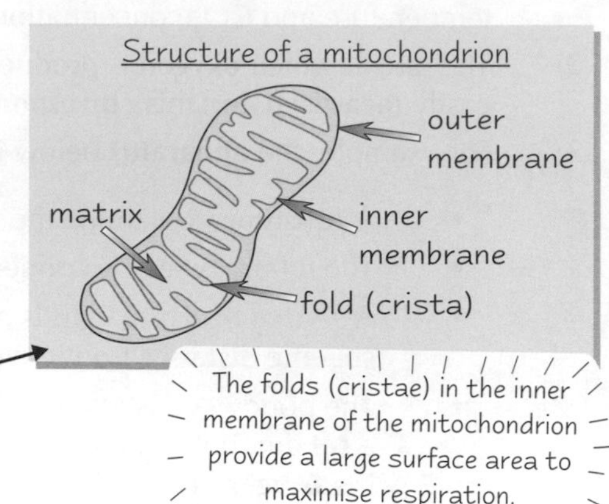

The folds (cristae) in the inner membrane of the mitochondrion provide a large surface area to maximise respiration.

Stage 1 — **Glycolysis** Makes **Pyruvate** from **Glucose**

1) Glycolysis involves splitting **one molecule** of glucose (with 6 carbons — 6C) into **two** smaller molecules of **pyruvate** (3C).
2) The process happens in the **cytoplasm** of cells.
3) Glycolysis is the **first stage** of both aerobic and anaerobic respiration and **doesn't need oxygen** to take place — so it's an **anaerobic** process.

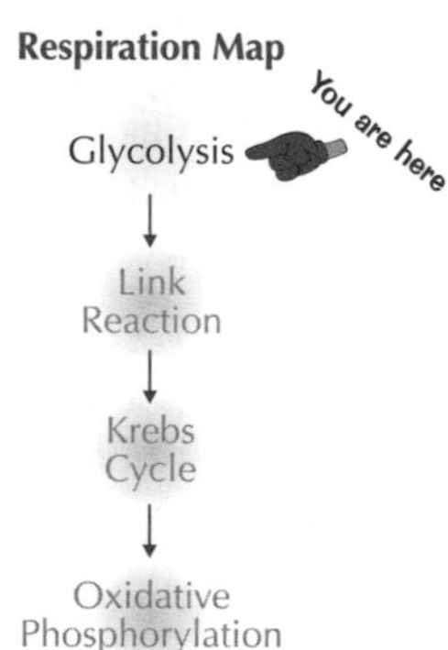

There are **Two Stages** in **Glycolysis** — **Phosphorylation** and **Oxidation**

First, **ATP** is **used** to **phosphorylate glucose** to triose phosphate. Then **triose phosphate** is **oxidised**, **releasing ATP**. Overall there's a **net gain** of **2 ATP**.

1) Stage One — Phosphorylation

1) Glucose is **phosphorylated** by adding **2 phosphates** from **2** molecules of **ATP**.
2) This creates **1** molecule of **hexose bisphosphate** and **2** molecules of **ADP**.
3) Then, **hexose bisphosphate** is **split** up into **2** molecules of **triose phosphate**.

2) Stage Two — Oxidation

1) Triose phosphate is **oxidised** (loses hydrogen), forming **2** molecules of **pyruvate**.
2) **NAD** collects the hydrogen ions, forming **2 reduced NAD**.
3) **4 ATP** are **produced**, but 2 were used up in stage one, so there's a **net gain** of **2 ATP**.

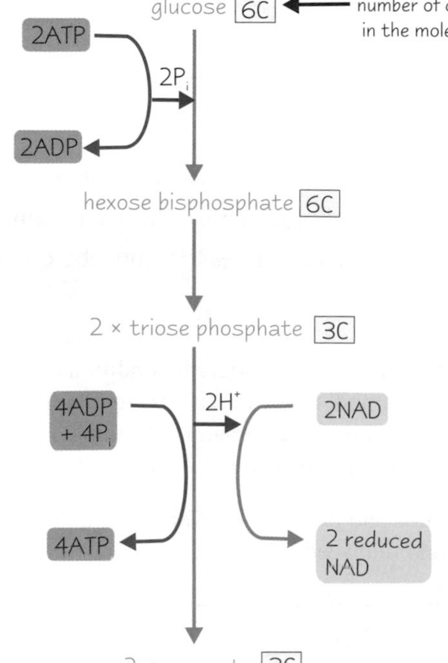

You're probably wondering what now happens to all the products of glycolysis...

1) The **two** molecules of **reduced NAD** go to the **last stage** (oxidative phosphorylation — see page 40).
2) The **two pyruvate** molecules are **actively transported** into the **matrix** of the **mitochondria** for the **link reaction** (see the next page).

Aerobic Respiration

Stage 2 — the Link Reaction converts Pyruvate to Acetyl Coenzyme A

The **link reaction** takes place in the **mitochondrial matrix**:

1) **Pyruvate** is **decarboxylated — one carbon atom** is **removed** from pyruvate in the form of $\mathbf{CO_2}$.
2) **NAD** is **reduced** — it collects **hydrogen** from pyruvate, changing **pyruvate** into **acetate**.
3) **Acetate** is combined with **coenzyme A** (CoA) to form **acetyl coenzyme A** (**acetyl CoA**).
4) **No ATP** is produced in this reaction.

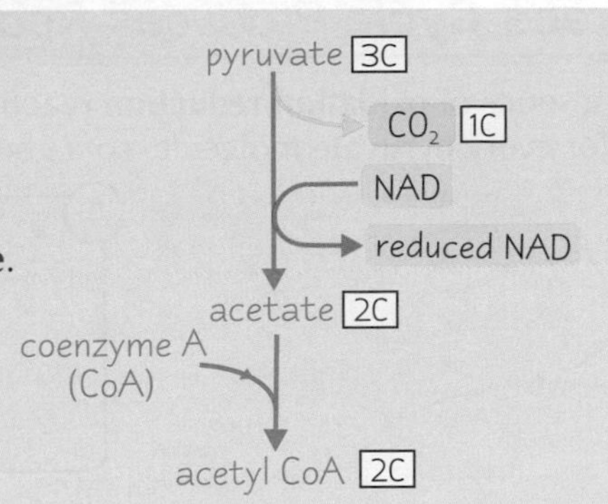

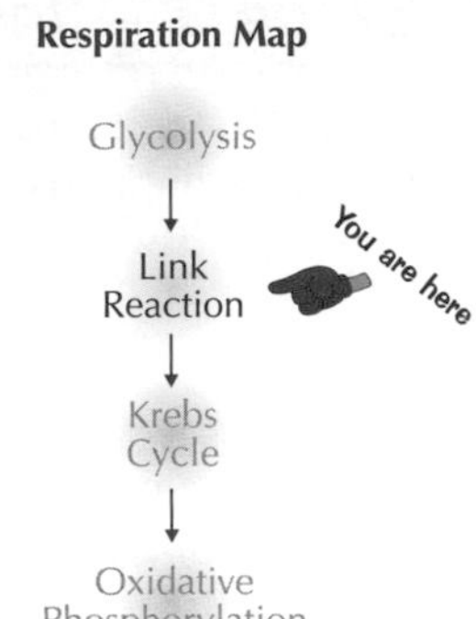

The Link Reaction occurs Twice for every Glucose Molecule

Two pyruvate molecules are made for **every glucose molecule** that enters glycolysis. This means the **link reaction** and the third stage (the **Krebs cycle**) happen **twice** for every glucose molecule. So for each glucose molecule:

- **Two** molecules of **acetyl coenzyme A** go into the Krebs cycle (see page 40).
- **Two** $\mathbf{CO_2}$ **molecules** are released as a waste product of respiration.
- **Two** molecules of **reduced NAD** are formed and go to the last stage (oxidative phosphorylation, see page 40).

Mitochondria are Adapted for Respiration

Mitochondria are **adapted** to their **function** in the following ways:

1) The **inner membrane** is **folded** into **cristae**, which **increases** the membrane's **surface area** to **maximise respiration**.
2) There are **lots** of **ATP synthase** molecules in the **inner membrane** to produce **lots** of **ATP** in the **final stage** of respiration.
3) The **matrix** contains all the **reactants** and **enzymes** needed for the **Krebs cycle** to take place.

Practice Questions

Q1 Where in the cell does glycolysis occur?

Q2 Is glycolysis an anaerobic or aerobic process?

Q3 How many ATP molecules are used up in glycolysis?

Q4 What is the product of the link reaction?

Exam Questions

Q1 Describe how a 6-carbon molecule of glucose is converted to pyruvate. [6 marks]

Q2 The link reaction of respiration occurs in the mitochondrial matrix.

b) Describe what happens in the link reaction. [3 marks]

a) Explain two ways in which a mitochondrion is adapted to its function. [2 marks]

No ATP was harmed during this reaction...

Ahhhh... too many reactions. I'm sure your head hurts now, 'cause mine certainly does. Just think of revision as like doing exercise — it can be a pain while you're doing it (and maybe afterwards too), but it's worth it for the well-toned brain you'll have. Just keep going over and over it, until you get the first two stages of respiration straight in your head. Then relax.

Aerobic Respiration

As you've seen, glycolysis produces a net gain of two ATP. Pah, we can do better than that. The Krebs cycle and oxidative phosphorylation are where it all happens — ATP galore.

Stage 3 — the Krebs Cycle Produces Reduced Coenzymes and ATP

The Krebs cycle involves a series of **oxidation-reduction reactions**, which take place in the **matrix** of the **mitochondria**. The cycle happens **once** for **every pyruvate** molecule, so it goes round **twice** for **every glucose** molecule.

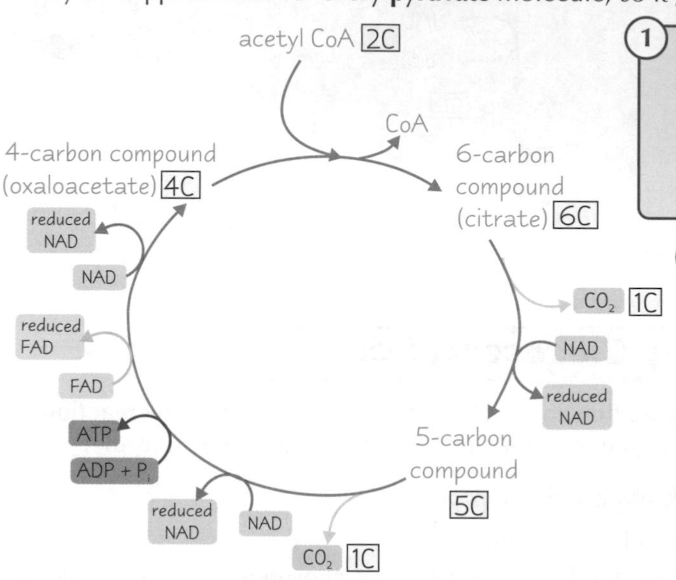

1
- **Acetyl CoA** from the link reaction combines with **oxaloacetate** to form **citrate**.
- **Coenzyme A** goes back to the **link reaction** to be used again.

2
- The **6C citrate molecule** is converted to a **5C molecule**.
- **Decarboxylation** occurs, where CO_2 **is removed**.
- **Dehydrogenation** also occurs, where **hydrogen** is **removed**.
- The hydrogen is used to **produce reduced NAD** from NAD.

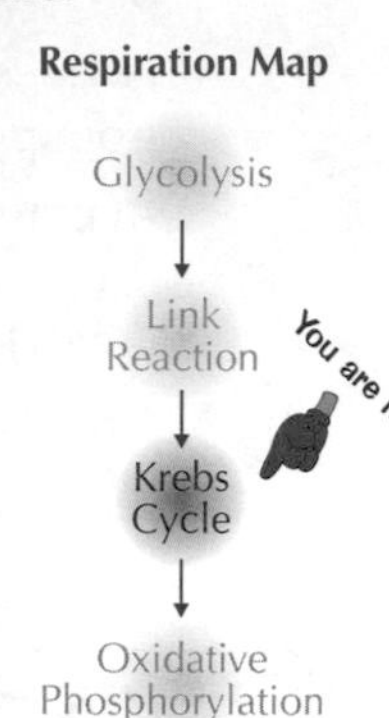

3
- The **5C molecule** is then converted to a **4C molecule**. (There are some intermediate compounds formed during this conversion, but you don't need to know about them.)
- **Decarboxylation** and **dehydrogenation** occur, producing **one** molecule of **reduced FAD** and **two** of **reduced NAD**.
- **ATP** is **produced** by the **direct transfer** of a **phosphate** group from an **intermediate** compound **to ADP**. When a phosphate group is directly transferred from one molecule to another it's called **substrate-level phosphorylation**. **Citrate** has now been **converted** into **oxaloacetate**.

Some Products of the Krebs Cycle are Used in Oxidative Phosphorylation

Some products are **reused**, some are **released** and others are used for the **next stage** of respiration:

Product from one Krebs cycle	Where it goes
1 coenzyme A	Reused in the next link reaction
Oxaloacetate	Regenerated for use in the next Krebs cycle
2 CO_2	Released as a waste product
1 ATP	Used for energy
3 reduced NAD	To oxidative phosphorylation
1 reduced FAD	To oxidative phosphorylation

Talking about oxidative phosphorylation was always a big hit with the ladies...

Stage 4 — Oxidative Phosphorylation Produces Lots of ATP

1) Oxidative phosphorylation is the process where the **energy** carried by **electrons**, from **reduced coenzymes** (reduced NAD and reduced FAD), is used to **make ATP**. (The whole point of the previous stages is to make reduced NAD and reduced FAD for the final stage.)
2) Oxidative phosphorylation involves two processes — the **electron transport chain** and **chemiosmosis** (see the next page).

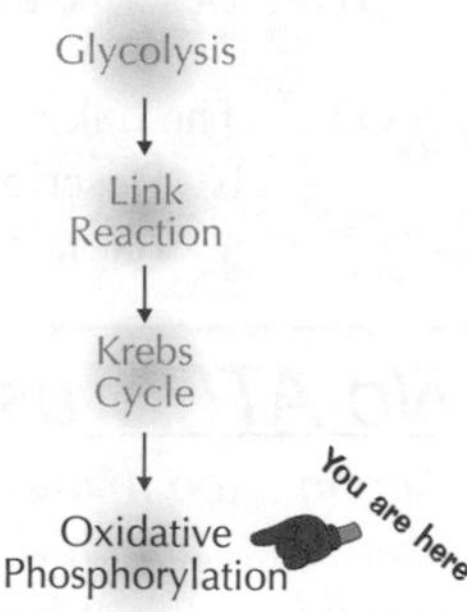

Aerobic Respiration

Protons are Pumped Across the Inner Mitochondrial Membrane

So now on to how **oxidative phosphorylation** actually **works**:

1) **Hydrogen atoms** are released from **reduced NAD** and **reduced FAD** as they're oxidised to NAD and FAD. The H atoms **split** into **protons (H^+)** and **electrons (e^-)**.
2) The **electrons** move along the **electron transport chain** (made up of three **electron carriers**), **losing energy** at each carrier.
3) This energy is used by the electron carriers to **pump protons** from the **mitochondrial matrix into** the **intermembrane space** (the space **between** the inner and outer **mitochondrial membranes**).
4) The **concentration** of **protons** is now **higher** in the **intermembrane space** than in the mitochondrial matrix — this forms an **electrochemical gradient** (a **concentration gradient** of **ions**).
5) Protons **move down** the **electrochemical gradient**, back into the mitochondrial matrix, via **ATP synthase**. This **movement** drives the synthesis of **ATP** from **ADP** and **inorganic phosphate** (P_i).
6) The movement of H^+ ions across a membrane, which generates ATP, is called **chemiosmosis**.
7) In the mitochondrial matrix, at the end of the transport chain, the **protons**, **electrons** and **O_2** (from the blood) combine to form **water**. Oxygen is said to be the final **electron acceptor**.

The regenerated coenzymes are reused in the Krebs cycle.

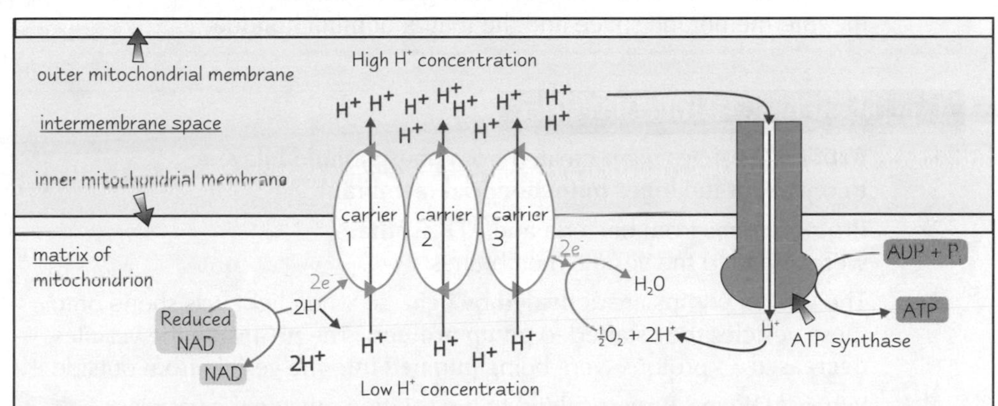

32 ATP Can be Made from One Glucose Molecule

As you know, **oxidative phosphorylation makes ATP** using energy from the reduced coenzymes — **2.5 ATP** are made from each **reduced NAD** and **1.5 ATP** are made from each **reduced FAD**. The table on the right shows **how much** ATP a cell can make from **one molecule** of **glucose** in **aerobic respiration.** (Remember, one molecule of glucose produces 2 pyruvate, so the link reaction and Krebs cycle happen twice.)

Stage of respiration	Molecules produced	Number of ATP molecules
Glycolysis	2 ATP	2
Glycolysis	2 reduced NAD	2 × 2.5 = 5
Link Reaction (×2)	2 reduced NAD	2 × 2.5 = 5
Krebs cycle (×2)	2 ATP	2
Krebs cycle (×2)	6 reduced NAD	6 × 2.5 = 15
Krebs cycle (×2)	2 reduced FAD	2 × 1.5 = 3
		Total ATP = 32

The number of ATP produced per reduced NAD or reduced FAD was thought to be 3 and 2, but new research has shown that the figures are nearer 2.5 and 1.5.

Practice Questions

Q1 Where in the cell does the Krebs cycle occur?

Q2 How many times does decarboxylation happen during one turn of the Krebs cycle?

Q3 What do the electrons lose as they move along the electron transport chain in oxidative phosphorylation?

Exam Question

Q1 Carbon monoxide inhibits the final electron carrier in the electron transport chain.
a) Explain how this affects ATP production via the electron transport chain. [2 marks]
b) Explain how this affects ATP production via the Krebs cycle. [2 marks]

The electron transport chain isn't just a FAD with the examiners...

Oh my gosh, I didn't think it could get any worse... You may be wondering how to learn these pages of crazy chemistry, but basically you have to put in the time and go over and over it. Don't worry though, it WILL pay off, and before you know it you'll be set for the exam. And once you know this section you'll be able to do anything, e.g. world domination...

Respiration Experiments

Congratulations — you've done all the main reactiony bits of respiration, so now it's time for some exciting experiments. When I say 'exciting', I'm using the word loosely. But I've got to say something positive to keep the morale up.

Scientific Experiments** Provide **Evidence** for **Chemiosmosis

Before the 1960s, scientists **didn't** understand the **connection** between the **electron transport chain** and **ATP synthesis** in respiration. One idea was that **energy lost** from **electrons moving** down the **electron transport chain** creates a **proton gradient** (a concentration gradient of H^+ ions), which is then used to **synthesise ATP** — this is called the **chemiosmotic theory**. Nowadays, there's quite a lot of **experimental evidence** supporting this theory:

Experiment One — Low pH

1) The **pH** of the **intermembrane space** in mitochondria was found to be **lower** than the pH of the **matrix**.
2) A **lower pH** means the intermembrane space is **more acidic** — it has a **higher concentration** of **H^+ ions**.
3) This observation shows that a **proton gradient exists** between the intermembrane space and the matrix of mitochondria.

The chemiosmotic theory is the most widely accepted theory for linking the electron transport chain to ATP synthesis.

Experiment Two — Artificial Vesicles

1) **Artificial vesicles** were created from **phospholipid bilayers** to **represent** the **inner mitochondrial membrane**.
2) **Proton pumps** from bacteria and **ATP synthase** were added to the vesicle membranes.
3) The **proton pumps** are **activated** by **light**, so when light was shone onto these vesicles they started to **pump protons**. The **pH inside** the vesicles **decreased** — protons were being **pumped into** the vesicle from outside.
4) When **ADP** and **P_i** were **added** to the solution **outside** the vesicles, **ATP** was **synthesised**.
5) This artificial system shows that a **proton gradient** can be **used** to **synthesise ATP** (but doesn't show that this happens in mitochondria).

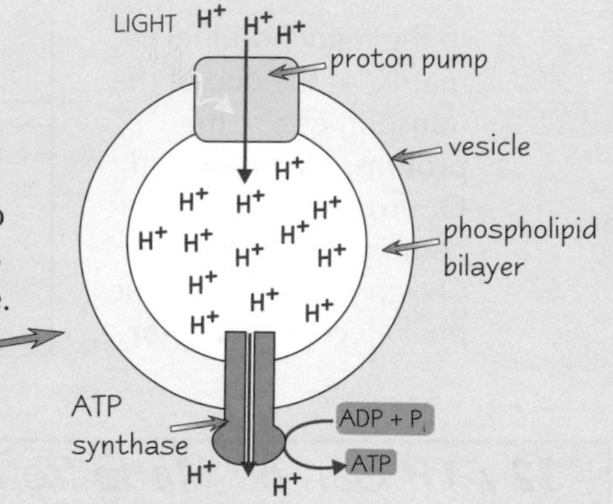

Experiment Three — Mitochondria

1) Mitochondria were put into a **slightly alkaline solution** (**pH8**).
2) They were left until the **whole** of each mitochondrion (matrix and intermembrane space) **became pH8**.
3) When these mitochondria were given **ADP** and **P_i** **no ATP** was produced.
4) Then the mitochondria were **transferred** to a **more acidic solution** of **pH4** (i.e. one with a **higher concentration** of **protons**).
5) The **outer membrane** of the mitochondrion is **permeable** to **protons** — the protons **moved into** the **intermembrane space**, creating a **proton gradient** across the **inner mitochondrial membrane**.
6) In the presence of **ADP** and **P_i**, **ATP was produced**.
7) This experiment shows that a **proton gradient** can be **used** by mitochondria to **make ATP**.

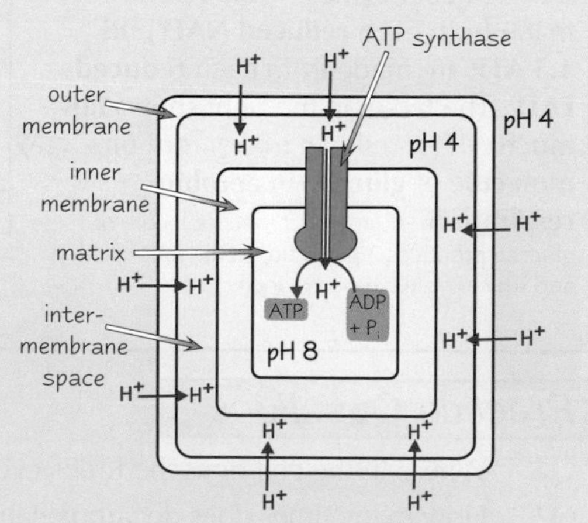

Experiment Four — Uncouplers

1) **Uncouplers** are substances that **destroy** the **proton gradient** across the **inner mitochondrial membrane**.
2) An **uncoupler** was added to mitochondria, along with **reduced NAD**, and **ADP** and **P_i**.
3) **No ATP** was made.
4) This experiment shows that a **proton gradient** is required to **synthesise ATP** in **mitochondria**.

Respiration Experiments

The Rate of Respiration can be Measured using a Respirometer

1) The volume of **oxygen taken up** or the volume of **carbon dioxide produced indicates** the **rate** of **respiration**.
2) A **respirometer** measures the rate of **oxygen** being **taken up** — the **more** oxygen taken up, the **faster** the rate of respiration.
3) Here's how you can use a **respirometer** to **measure** the volume of **oxygen taken up** by some **woodlice**:

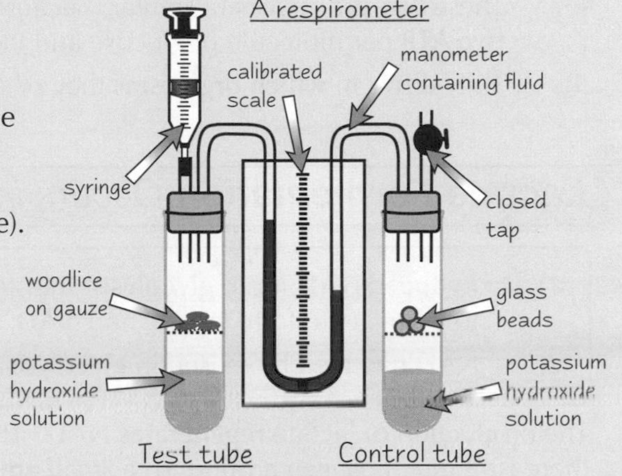

- The apparatus is set up as shown on the right.
- **Each tube** contains **potassium hydroxide** solution (or soda lime), which **absorbs carbon dioxide.**
- The **control tube** is set up in exactly the **same way** as the test tube, but **without** the **woodlice**, to make sure the **results** are **only** due to the woodlice **respiring** (e.g. it contains beads that have the same mass as the woodlice).
- The **syringe** is used to set the **fluid** in the **manometer** to a **known level.**
- The apparatus is **left** for a **set** period of **time** (e.g. 20 minutes).
- During that time there'll be a **decrease** in the **volume** of the **air** in the test tube, due to **oxygen consumption** by the **woodlice** (all the CO_2 produced is absorbed by the potassium hydroxide).
- The decrease in the volume of the air will **reduce the pressure** in the tube and cause the **coloured liquid** in the manometer to **move towards** the test tube.
- The **distance moved** by the **liquid** in a **given time** is **measured.** This value can then be used to **calculate** the **volume of oxygen** taken in by the woodlice **per minute.**
- Any **variables** that could **affect** the results are **controlled**, e.g. temperature, volume of potassium hydroxide solution in each test tube.

4) To produce more **reliable** results the experiment is **repeated** and a **mean volume** of O_2 is calculated.

Practice Questions

Q1 What is the chemiosmotic theory?

Q2 What does a respirometer measure?

Exam Questions

Q1 In the first stage of an experiment, mitochondria were put in a solution at pH 9.1 and left until each mitochondrion had a pH of 9.1 throughout its compartments. In the presence of ADP and P_i, no ATP was produced. During the second stage of the experiment, the same mitochondria were placed in a solution at pH 3.7. In the presence of ADP and P_i, ATP was produced.

a) Why was no ATP produced during the first part of the experiment? [1 mark]

b) What would the pH of the intermembrane space have been during the second stage of the experiment? [1 mark]

c) Do these results support the chemiosmotic theory? Explain your answer. [1 mark]

Q2 A respirometer is set up as shown in the diagram on this page.

a) Explain the purpose of the control tube. [1 mark]

b) Explain what would happen if there was no potassium hydroxide in the tubes. [2 marks]

c) What other substance could be measured to find out the rate of respiration? [1 mark]

My results are dodgy — I'm sure the woodlice are holding their breath...

Okay, that wasn't very funny, but this page doesn't really give me any inspiration. You probably feel the same way. It's just one of those pages that you have to plough through. You could try drawing a few pretty diagrams to get the experiments in your head. And after you've got it sorted do something exciting, like trying to stick your toe in your ear...

Aerobic and Anaerobic Respiration

We're on the home stretch now ladies and gents — these are the last two pages in the section.

There are Two Types of Anaerobic Respiration

1) **Anaerobic** respiration **doesn't use oxygen**.
2) It **doesn't** involve the **link reaction**, the **Krebs cycle** or **oxidative phosphorylation**.
3) There are **two types** of anaerobic respiration — **alcoholic fermentation** and **lactate fermentation**.
4) These two processes are **similar**, because they both take place in the **cytoplasm**, they both produce **two ATP** per molecule of glucose and they both **start** with **glycolysis** (which produces **pyruvate**).
5) They **differ** in **which organisms** they occur in and what happens to the **pyruvate** (see below).

Lactate Fermentation Occurs in Mammals and Produces Lactate

1) **Reduced NAD** (from glycolysis) transfers **hydrogen** to **pyruvate** to form **lactate** and **NAD**.
2) **NAD** can then be reused in **glycolysis**.

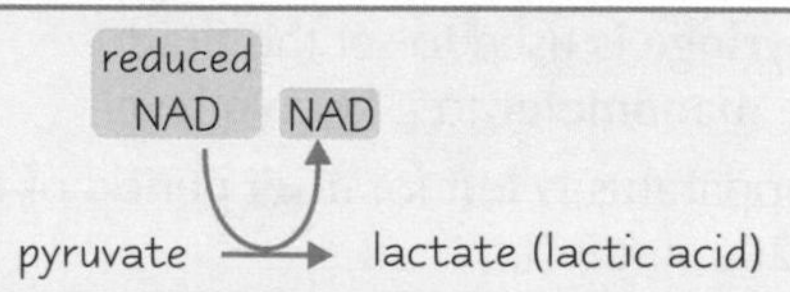

Some bacteria carry out lactate fermentation.

The production of lactate **regenerates NAD**. This means **glycolysis** can **continue** even when there **isn't** much oxygen around, so a **small amount of ATP** can still be **produced** to keep some biological process going... clever.

Alcoholic Fermentation Occurs in Yeast Cells and Produces Ethanol

1) CO_2 is **removed** from **pyruvate** to form **ethanal**.
2) **Reduced NAD** (from glycolysis) transfers **hydrogen** to **ethanal** to form **ethanol** and **NAD**.
3) **NAD** can then be reused in **glycolysis**.

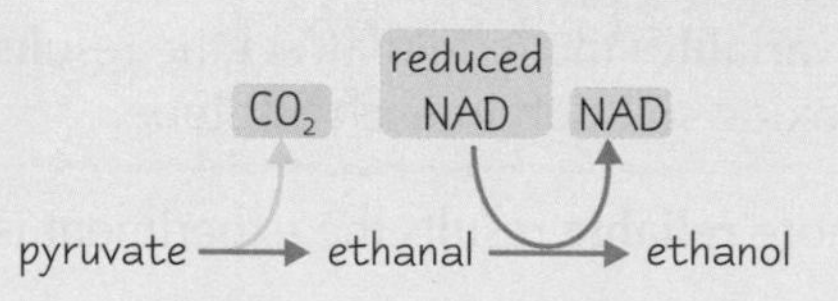

Alcoholic fermentation also occurs in plants.

The production of ethanol also **regenerates NAD** so **glycolysis** can **continue** when there isn't much oxygen around.

Aerobic Respiration Doesn't Release as Much Energy as Possible...

In theory, **aerobic respiration** can make **32 ATP** per **glucose molecule** (see page 41).
But in reality the **actual yield** is **lower** because:

1) Some of the **reduced NAD** formed during the **first three stages** of aerobic respiration is used in **other reduction reactions** in the cell instead of in **oxidative phosphorylation**.
2) **Some ATP** is **used up** by **actively transporting** substances **into** the **mitochondria** during respiration, e.g. **pyruvate** (formed at the end of glycolysis), **ADP** and **phosphate** (both needed for making ATP).
3) The **inner mitochondrial membrane** is **leaky** — some **protons** may **leak** into the **matrix** without passing through **ATP synthase** and **without making ATP**.

...but it Still Releases More Energy than Anaerobic Respiration

1) The **ATP yield** from **anaerobic** respiration is **always lower** than from **aerobic** respiration.
2) This is because **anaerobic** respiration **only** includes **one energy-releasing stage** (**glycolysis**), which only produces **2 ATP** per glucose molecule.
3) The energy-releasing reactions of the **Krebs cycle** and **oxidative phosphorylation** need **oxygen**, so they **can't** occur during anaerobic respiration.

Aerobic and Anaerobic Respiration

Cells Can Respire Different Substrates

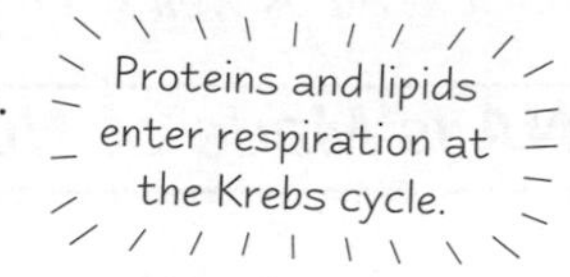

1) Cells **respire glucose**, but they also respire **other carbohydrates**, **lipids** and **proteins**.
2) Any **biological molecule** that can be **broken down** in **respiration** to **release energy** is called a **respiratory substrate**.
3) When an organism respires a specific **respiratory substrate**, the **respiratory quotient** (RQ) can be **worked out**.
4) The **respiratory quotient** is the volume of **carbon dioxide** produced when that **substrate** is **respired**, **divided** by the volume of **oxygen consumed**, in a set period of **time**.

$$RQ = \frac{\text{Volume of } CO_2 \text{ released}}{\text{Volume of } O_2 \text{ consumed}}$$

5) For example, you can work out the **RQ** for cells that **only respire glucose**:
   - The basic equation for aerobic respiration using glucose is: $C_6H_{12}O_6 + 6O_2 \rightarrow 6CO_2 + 6H_2O$ + energy
   - The RQ of glucose = molecules of **CO_2 released** ÷ molecules of **O_2 consumed** **= 6 ÷ 6 = 1**.
6) Respiratory quotients have been worked out for the respiration of **other respiratory substrates**. **Lipids** and **proteins** have an RQ value **lower than one** because **more oxygen** is needed to oxidise fats and lipids than to oxidise carbohydrates.

Respiratory Substrate	RQ
Lipids (triglycerides)	0.7
Proteins or amino acids	0.9
Carbohydrates	1

The Respiratory Quotient tells you what Substrate is being Respired

1) The **respiratory quotient** for an organism is **useful** because it tells you **what kind** of **respiratory substrate** an organism is respiring and what **type** of **respiration** it's using (aerobic or anaerobic).
2) For example, under **normal conditions** the **usual RQ** for humans is between **0.7** and **1.0**. An RQ in this range shows that some **fats** (**lipids**) are being used for respiration, as well as **carbohydrates** like glucose. Protein **isn't** normally used by the body for respiration unless there's **nothing else**.
3) **High RQs** (greater than 1) mean that an organism is **short** of **oxygen**, and is having to respire **anaerobically** as well as aerobically.
4) **Plants** sometimes have a **low RQ**. This is because the **CO_2 released** in respiration is **used** for **photosynthesis** (so it's not measured).

Practice Questions

Q1 What molecule is made when CO_2 is removed from pyruvate during alcoholic fermentation?

Q2 Does anaerobic respiration release more or less energy per glucose molecule than aerobic respiration?

Q3 What is a respiratory substrate?

Exam Questions

Q1 A culture of mammalian cells was incubated with glucose, pyruvate and antimycin C. Antimycin C inhibits an electron carrier in the electron transport chain of aerobic respiration. Explain why these cells can still produce lactate. [1 mark]

Q2 This equation shows the aerobic respiration of a fat called triolein: $C_{57}H_{104}O_6 + 80O_2 \rightarrow 52H_2O + 57CO_2$
Calculate the respiratory quotient for this reaction. Show your working. [2 marks]

My little sis has an RQ of 157 — she's really clever...

I know, I'm really pushing the boundary between humour and non-humour here. But, at least we've come to the end of the section — and what a section it was. You might think it's unfair finishing it off with nasty calculations, but if you understand how to work out the RQ you'll be one step closer to being sorted for the exam.

DNA, RNA and Protein Synthesis

You learnt how DNA and its mysterious cousin RNA are used to produce proteins at AS, but irritatingly you need to know it at A2 as well (with a few extra bits thrown in — unlucky).

DNA is Made of Nucleotides that Have a Sugar, a Phosphate and a Base

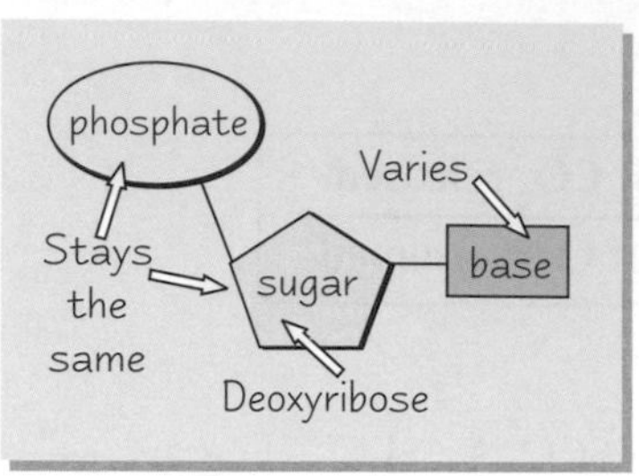

1) DNA is a **polynucleotide** — it's made up of lots of **nucleotides** joined together.
2) Each nucleotide is made from a **pentose sugar** (with 5 carbon atoms), a **phosphate** group and a **nitrogenous base**.
3) The **sugar** in DNA nucleotides is a **deoxyribose** sugar.
4) Each nucleotide has the **same sugar and phosphate**. The **base** on each nucleotide can **vary** though.
5) There are **four** possible bases — adenine (**A**), thymine (**T**), cytosine (**C**) and guanine (**G**).

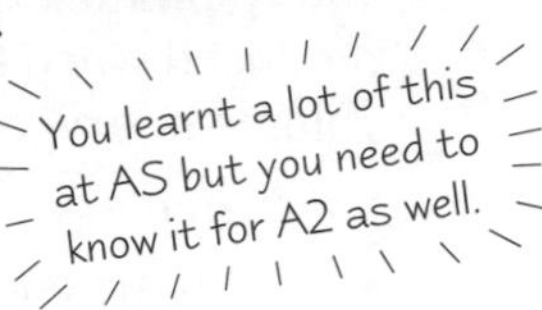

Two Polynucleotide Strands Join Together to Form a Double-Helix

1) DNA nucleotides join together to form **polynucleotide strands**.
2) The nucleotides join up between the **phosphate** group of one nucleotide and the **sugar** of another, creating a **sugar-phosphate backbone**.
3) **Two** DNA polynucleotide strands join together by **hydrogen bonding** between the bases.
4) Each base can only join with one particular partner — this is called **complementary base pairing**.
5) **Adenine** always pairs with **thymine** (**A - T**) and **guanine** always pairs with **cytosine** (**G - C**).
6) The two strands **wind up** to form the **DNA double-helix**.

When two strands have bases that pair up the strands are said to be complementary to each other:

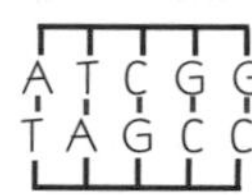

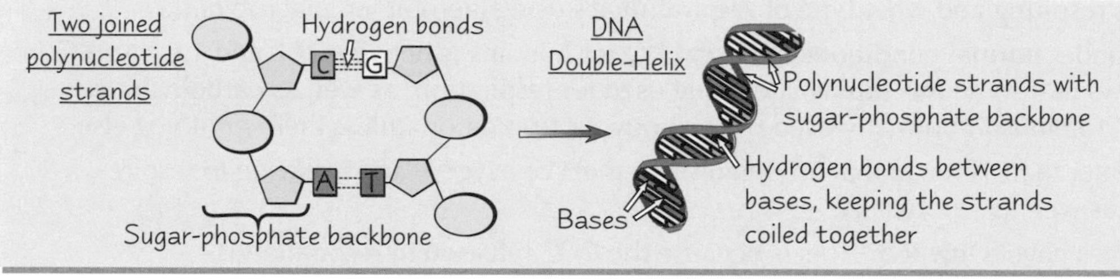

DNA Contains Genes Which are Instructions for Proteins

Polypeptide is just another word for a protein.

1) Genes are **sections of DNA**. They're found on **chromosomes**.
2) Genes **code** for **proteins** (polypeptides), including **enzymes** — they contain the **instructions** to make them.
3) Proteins are made from **amino acids**. Different proteins have a **different number** and **order** of amino acids.
4) It's the **order** of **bases** in a gene that determines the **order of amino acids** in a particular **protein**.
5) Each amino acid is coded for by a sequence of **three bases** (called a **triplet** or a **codon**) in a gene.
6) **Different sequences** of bases code for different amino acids — this is the **genetic code**.

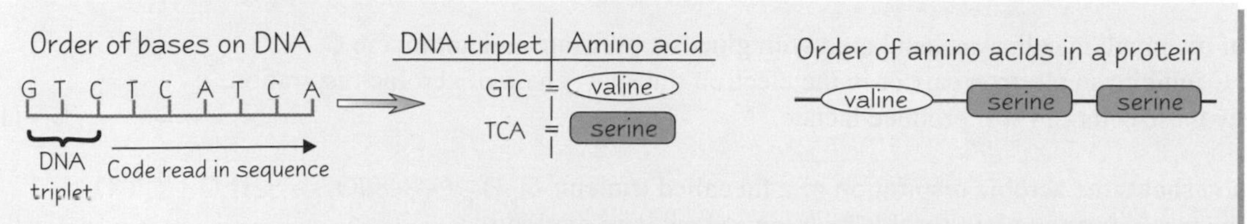

7) Some amino acids are coded for by **more than one** triplet, e.g. CGA, CGG, CGT and CGC all code for arginine.
8) Other triplets are used to tell the cell when to **start** and **stop** production of the protein — these are called **start** and **stop codons**. They're found at the **beginning** and **end** of the gene. E.g. TAG is a stop codon.

Stop codons are also called stop signals.

DNA, RNA and Protein Synthesis

DNA is Copied into RNA for Protein Synthesis

1) DNA molecules are found in the **nucleus** of the cell, but the organelles for protein synthesis (**ribosomes**) are found in the **cytoplasm**.
2) DNA is too large to move out of the nucleus, so a section is **copied** into **RNA**. This process is called **transcription** (see next page).
3) RNA is a **single** polynucleotide strand — it contains the sugar **ribose**, and **uracil** (**U**) replaces thymine as a base. Uracil **always pairs** with **adenine** during protein synthesis.
4) The RNA **leaves** the nucleus and joins with a **ribosome** in the cytoplasm, where it can be used to synthesise a **protein**. This process is called **translation** (see page 49).
5) There are actually two types of RNA you need to know about:

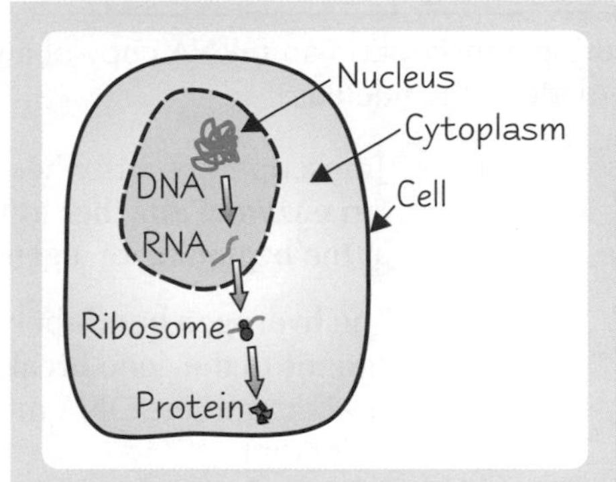

Messenger RNA (mRNA)
- Made in the **nucleus**.
- **Three adjacent bases** are called a **codon**.
- It **carries the genetic code** from the DNA in the **nucleus** to the **cytoplasm**, where it's used to make a **protein** during **translation**.

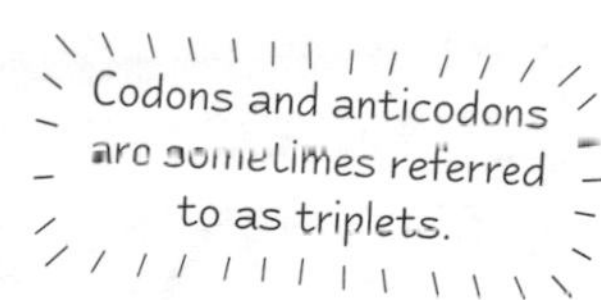

Transfer RNA (tRNA)
- Found in the **cytoplasm**.
- It has an **amino acid binding site** at one end and a **sequence** of **three bases** at the other end called an **anticodon**.
- It **carries** the **amino acids** that are used to make **proteins** to the **ribosomes** during **translation**.

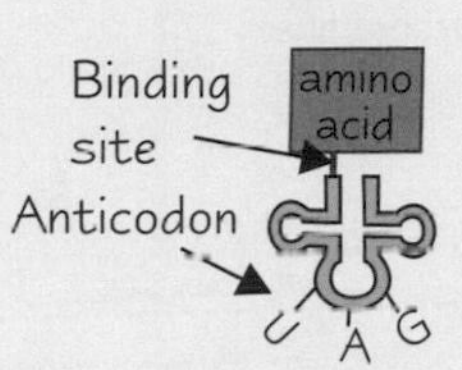

Practice Questions

Q1 Name the four possible bases in DNA.

Q2 Which DNA bases pair together in complementary base pairing?

Q3 What determines the order of amino acids in a protein?

Q4 Name the bases found in RNA.

Q5 What are three adjacent bases in mRNA called?

Q6 Where in the cell is mRNA made?

Amino Acid	DNA sequence
Serine	AGA
Leucine	GAT
Tyrosine	ATA
Valine	CAC
Alanine	CGT
Arginine	GCG

Exam Questions

Q1 Describe the role of mRNA and the role of tRNA. [2 marks]

Q2 A piece of DNA has the following nucleotide sequence: AGAAGAATACACCGT
a) How many amino acids does this sequence code for? [1 mark]
b) Using the table above, write down the amino acid sequence it codes for. [2 marks]

Genes, genes are good for your heart, the more you eat, the more you...

Hurrah — finally some pages with something familiar on. But just because you learnt some of it at AS doesn't mean you can skip them. You really need to get your head around how DNA and RNA work together to produce proteins or the next two pages are going to be a teeeny weeny bit tricky. Don't say I didn't warn you. Turn over too quickly at your own peril...

Transcription and Translation

Time to find out how RNA works its magic to make proteins. It gets a bit complicated but bear with it.

First Stage of Protein Synthesis — Transcription

During transcription an **mRNA copy** of a gene (a section of DNA) is made in the **nucleus**:

1) Transcription starts when **RNA polymerase** (an **enzyme**) **attaches** to the **DNA** double-helix at the **beginning** of a **gene**.
2) The **hydrogen bonds** between the two DNA strands in the gene **break**, **separating** the strands, and the DNA molecule **uncoils** at that point.
3) One of the strands is then used as a **template** to make an **mRNA copy**.

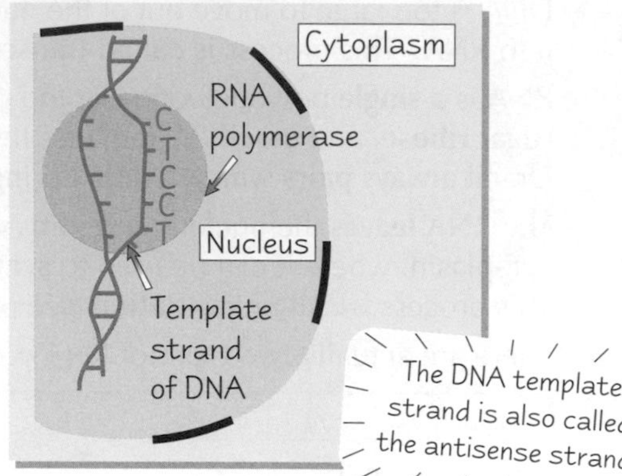

The DNA template strand is also called the antisense strand.

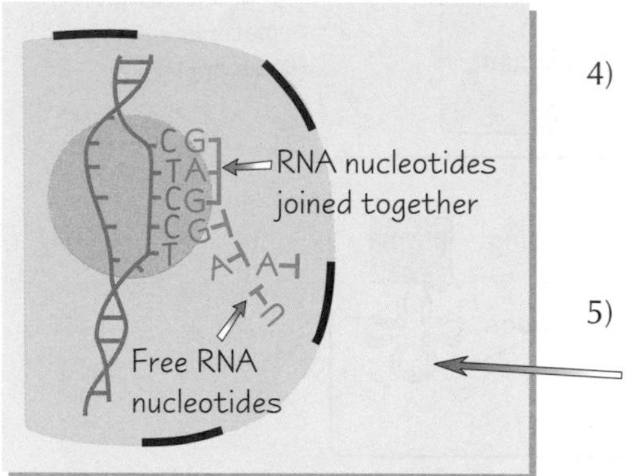

4) The RNA polymerase **lines up** free **RNA nucleotides** alongside the template strand. **Complementary base pairing** means that the mRNA strand ends up being a **complementary copy** of the DNA template strand (except the base **T** is replaced by **U** in **RNA**).
5) Once the RNA nucleotides have **paired up** with their **specific bases** on the DNA strand they're **joined together**, forming an **mRNA** molecule.

6) The RNA polymerase moves **along** the DNA, separating the strands and **assembling** the mRNA strand.
7) The **hydrogen bonds** between the uncoiled strands of DNA **re-form** once the RNA polymerase has passed by and the strands **coil back into a double-helix**.

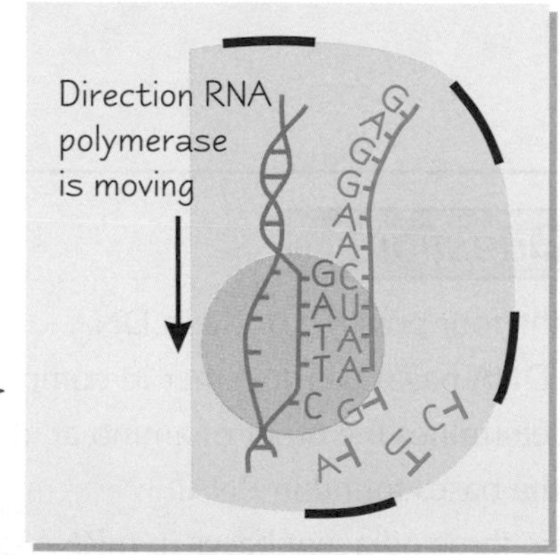

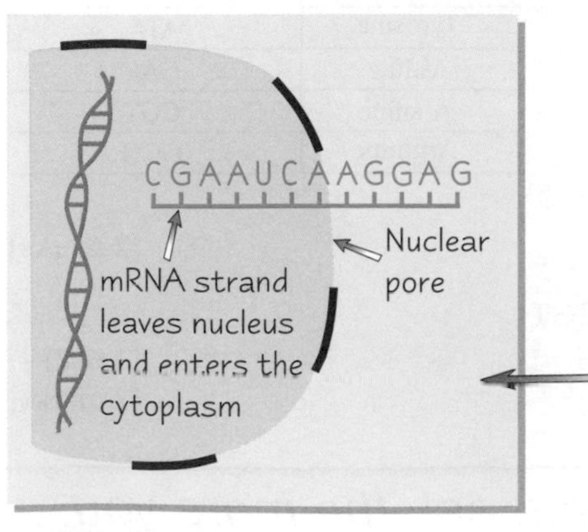

8) When RNA polymerase reaches a **stop codon**, it stops making mRNA and **detaches** from the DNA.
9) The **mRNA** moves **out** of the **nucleus** through a **nuclear pore** and attaches to a **ribosome** in the cytoplasm, where the next stage of protein synthesis takes place (see next page).

Transcription and Translation

Second Stage of Protein Synthesis — Translation

Translation occurs at the **ribosomes** in the **cytoplasm**. During **translation**, **amino acids** are **joined together** to make a **polypeptide chain** (protein), following the sequence of **codons** carried by the mRNA.

1) The **mRNA attaches** itself to a **ribosome** and **transfer RNA** (**tRNA**) molecules **carry amino acids** to the ribosome.
2) A tRNA molecule, with an **anticodon** that's **complementary** to the **first codon** on the mRNA, attaches itself to the mRNA by **complementary base pairing**.
3) A second tRNA molecule attaches itself to the **next codon** on the mRNA in the **same way**.
4) The two amino acids attached to the tRNA molecules are **joined** by a **peptide bond**. The first tRNA molecule **moves away**, leaving its amino acid behind.
5) A third tRNA molecule binds to the **next codon** on the mRNA. Its amino acid **binds** to the first two and the second tRNA molecule **moves away**.
6) This process continues, producing a chain of linked amino acids (a **polypeptide chain**), until there's a **stop codon** on the mRNA molecule.
7) The polypeptide chain (**protein**) **moves away** from the ribosome and translation is complete.

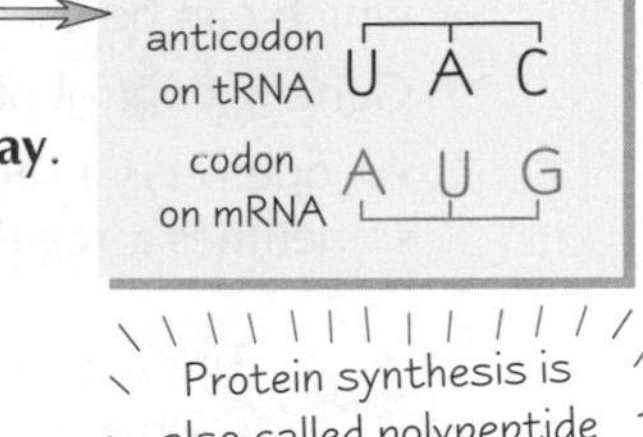

Protein synthesis is also called polypeptide synthesis as it makes a polypeptide (protein)

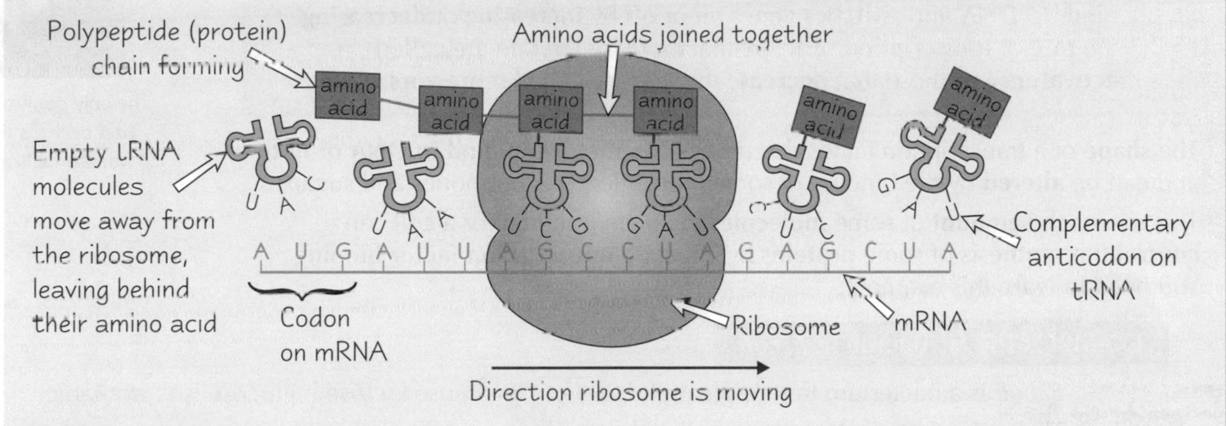

Practice Questions

Q1 What are the two stages of protein synthesis called?

Q2 Where does the first stage of protein synthesis take place?

Q3 When does RNA polymerase stop making mRNA?

Q4 Where does the second stage of protein synthesis take place?

Exam Questions

Q1 A drug that inhibits cell growth is found to be able to bind to DNA, preventing RNA polymerase from binding. Explain how this drug will affect protein synthesis. [2 marks]

Q2 A polypeptide chain (protein) from a eukaryotic cell is 10 amino acids long.
a) Predict how long the mRNA for this protein would be in nucleotides (without the start and stop codons). Explain your answer. [2 marks]
b) Describe how the mRNA is translated into the polypeptide chain. [6 marks]

The only translation I'm interested in is a translation of this page into English...

So you start off with DNA, lots of cleverness happens and bingo... you've got a protein. Only problem is you need to know the cleverness bit in quite a lot of detail. So scribble it down, recite it to yourself, explain it to your best mate or do whatever else helps you remember the joys of protein synthesis. And then think how clever you must be to know it all.

Control of Protein Synthesis and Body Plans

Proteins aren't just made willy-nilly — there's some control over when they're synthesised...

Genes can be **Switched On** or **Off**

1) **Protein synthesis** can be **controlled** at the **genetic level** by **altering** the rate of **transcription** of genes. E.g. **increased** transcription produces **more mRNA**, which can be used to make **more protein**.

2) Genetic control of protein production in **prokaryotes** (e.g. bacteria) often involves **operons**.
3) An operon is a **section** of **DNA** that contains **structural genes**, **control elements** and sometimes a **regulatory gene**:

- The structural genes code for **useful proteins**, such as **enzymes** — they're all **transcribed together**.
- The control elements include a **promoter** (a DNA sequence located **before** the structural genes that **RNA polymerase** binds to) and an **operator** (a DNA sequence that proteins called **transcription factors** bind to).
- The regulatory gene codes for a **transcription factor** — a protein that **binds to DNA** and **switches** genes **on** or **off** by **increasing** or **decreasing** the **rate** of **transcription**. Factors that **increase** the rate are called **activators** and those that **decrease** the rate are called **repressors**.

The only control Brad had over his jeans was some braces.

4) The **shape** of a transcription factor determines whether it **can bind to DNA** or **not**, and can be **altered** by the binding of some molecules, e.g. hormones and sugars.
5) This means the **amount** of some **molecules** in an environment or a cell can **control** the **synthesis** of some **proteins** by affecting **transcription factor binding**. You need to learn this example:

EXAMPLE: The *lac* operon in *E. coli*

1) ***E. coli*** is a bacterium that **respires glucose**, but it can use **lactose** if glucose isn't available.
2) The genes that produce the **enzymes** needed to **respire lactose** are found on an operon called the **lac operon**.
3) The lac operon has **three structural genes** — **lacZ**, **lacY** and **lacA**, which produce proteins that help the bacteria digest lactose (including **β-galactosidase** and **lactose permease**).
4) Here's how it works:

Lactose NOT present

The **regulatory** gene (lacI) produces the **lac repressor**, which **binds** to the **operator** site when there's **no lactose** present and **blocks transcription**.

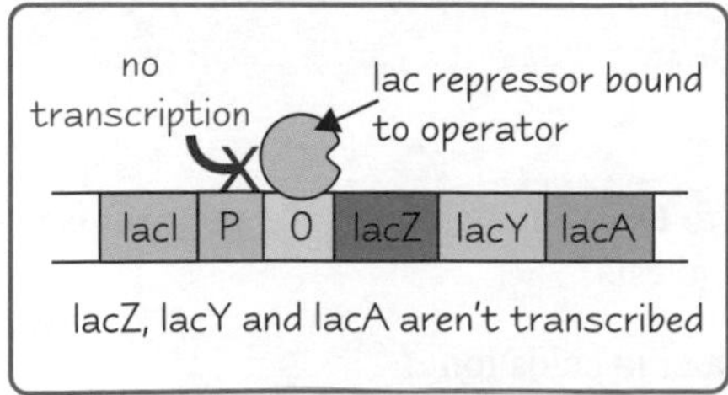

Lactose present

When **lactose is present**, it **binds** to the **repressor**, **changing** the repressor's **shape** so that it can **no longer bind** to the operator site.
RNA polymerase can now **begin transcription** of the structural genes.

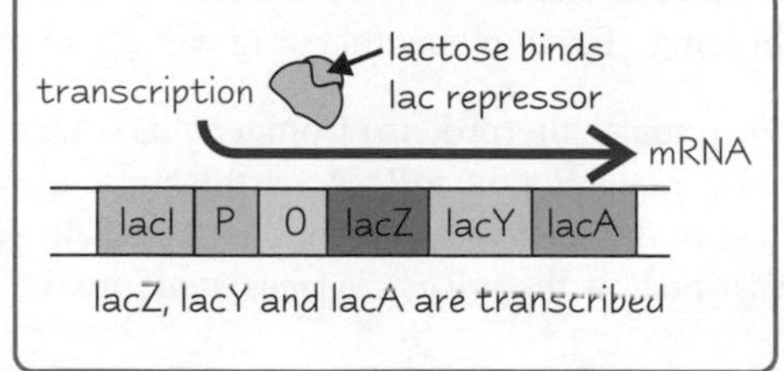

Control of Protein Synthesis and Body Plans

Some Genes Control the Development of Body Plans

1) A **body plan** is the **general structure** of an organism, e.g. the *Drosophila* fruit fly has various **body parts** (head, abdomen, etc.) that are **arranged** in a **particular way** — this is its body plan.
2) **Proteins control** the **development** of a **body plan** — they help set up the basic body plan so that everything is in the right place, e.g. legs grow where legs should grow.
3) The proteins that control body plan development are **coded for** by genes called **homeotic genes**. E.g. two homeotic gene clusters control the development of the *Drosophila* body plan — one controls the development of the head and anterior thorax and the other controls the development of the posterior thorax and abdomen.

anterior thorax | posterior thorax | head | abdomen

4) **Similar homeotic genes** are found in **animals**, **plants** and **fungi**, which means that body plan development is controlled in a similar way in flies, mice, humans, etc. Here's how homeotic genes control development:
   - **Homeotic genes** have **regions** called **homeobox sequences** that code for a part of the protein called the **homeodomain**.
   - The homeodomain **binds** to specific **sites** on **DNA**, enabling the protein to work as a **transcription factor** (see previous page).
   - The proteins bind to DNA at the **start** of **developmental genes**, **activating** or **repressing transcription** and so altering the production of proteins involved in the development of the body plan.

Programmed Cell Death is Involved in the Development of Body Plans

1) Some cells **die** and **break down** as a **normal** part of **development**.
2) This is a **highly controlled process** called **apoptosis**, or **programmed cell death**.
3) Once **apoptosis** has been **triggered** the **cell** is **broken down** in a series of steps:
   - The cell produces **enzymes** that **break down** important cell components such as **proteins** in the cytoplasm and **DNA** in the nucleus.
   - As the cell's contents are broken down it begins to **shrink** and **breaks up** into **fragments**.
   - The **cell fragments** are **engulfed** by **phagocytes** and **digested**.
4) Apoptosis is involved in the development of **body plans** — **mitosis** and **differentiation create** the bulk of the **body parts** and then apoptosis **refines** the parts by **removing** the **unwanted structures**. For example:
   - When **hands** and **feet** first develop the **digits** (fingers and toes) are **connected**. They're only **separated** when cells in the **connecting tissue** undergo **apoptosis**.
   - As **tadpoles** develop into frogs their **tail cells** are **removed** by apoptosis.
   - An **excess** of **nerve cells** are produced during the development of the **nervous system**. Nerve cells that **aren't needed** undergo **apoptosis**.
5) All cells contain **genes** that code for proteins that **promote** or **inhibit apoptosis**.
6) During development, genes that **control** apoptosis are **switched on** and **off** in **appropriate** cells, so that **some die** and the **correct body plan develops**.

Practice Questions

Q1 What does a transcription factor do?

Q2 What is a body plan?

Q3 Give one example of how apoptosis is used during the development of a body plan.

Exam Question

Q1 Explain how the presence of lactose causes *E. coli* to produce ß-galactosidase and lactose permease. [4 marks]

Too much revision can activate programmed cell death in your brain....

OK, maybe that's not completely true. There are a lot of 'interesting' words to remember on these two pages — I bet you're glad you decided to study biology. Some of these concepts are quite hard to get your head round but keep going over it until it all makes sense — it'll click eventually. Then you can dazzle your friends with your knowledge of gene control...

Protein Activation and Gene Mutation

Some proteins need activating before they'll work — cyclic AMP gives them a molecular kick up their amino acid backside. Proteins can also be affected by mutations in DNA... sounds like it's a hard life being a protein...

cAMP Activates Some Proteins by Altering Their Shape

1) **Protein synthesis** can be controlled at the **genetic level** by **molecules** (see page 50).
2) Some proteins produced by protein synthesis **aren't active** though — they have to be **activated** to work.
3) **Protein activation** is also controlled by **molecules**, e.g. **hormones** and **sugars**.
4) Some of these molecules work by **binding** to **cell membranes** and **triggering** the production of cyclic **AMP** (**cAMP**) **inside** the **cell**.
5) cAMP then **activates proteins** inside the cell by **altering** their **three-dimensional** (3D) **structure**.
6) For example, altering the 3D structure can **change** the **active site** of an enzyme, making it become **more** or **less active**.
7) For example, cAMP activates **protein kinase A** (**PKA**):

cAMP is a secondary messenger — it relays the message from the control molecule, e.g. the hormone, to the inside of the cell (see page 12).

1) **PKA** is an **enzyme** made of four subunits.
2) When cAMP **isn't bound**, the four units are bound together and are **inactive**.
3) When cAMP **binds**, it causes a **change** in the enzyme's **3D structure**, releasing the active subunits — PKA is now **active**.

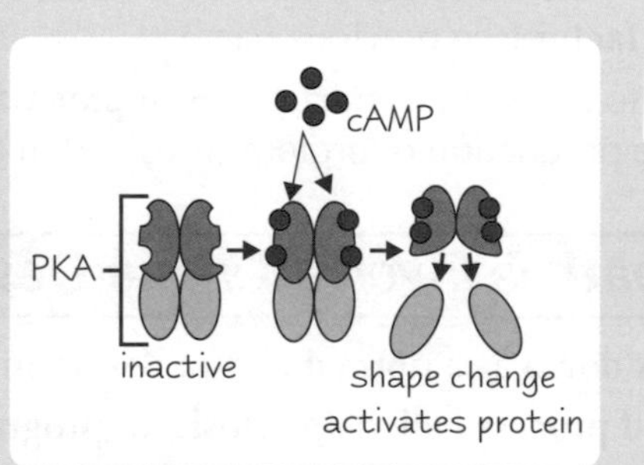

Mutations are Changes to the Base Sequence of DNA

1) Any change to the **base (nucleotide) sequence** of DNA is called a **mutation**.
2) The **types** of mutations that can occur include:

- **Substitution** — one base is swapped for another, e.g. ATGCCT becomes ATTCCT
- **Deletion** — one base is removed, e.g. ATGCCT becomes ATCCT
- **Insertion** — one base is added, e.g. ATGCCT becomes ATGACCT
- **Duplication** — one or more bases are repeated, e.g. ATGCCT becomes ATGCCCCT
- **Inversion** — a sequence of bases is reversed, e.g. ATGCCT becomes ACCGTT

3) The **order** of **DNA bases** in a gene determines the **order of amino acids** in a particular **protein**. If a mutation occurs in a gene, the **primary structure** (amino acid chain) of the protein it codes for could be **altered**:

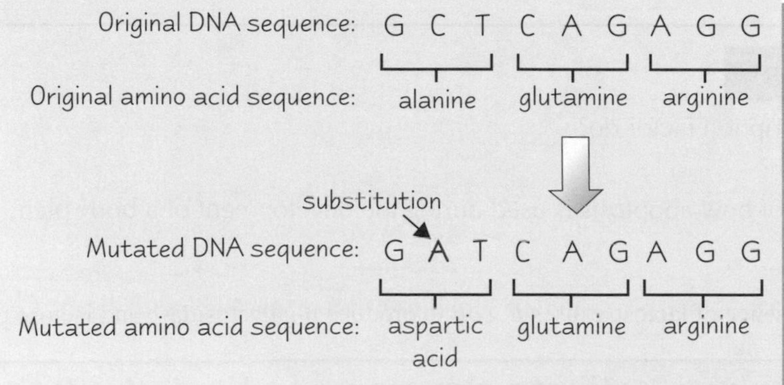

4) This may **change** the final **3D shape** of the protein so it **doesn't work properly**, e.g. **active sites** in enzymes may not form properly, meaning that **substrates can't bind** to them.

Protein Activation and Gene Mutation

Mutations can be **Neutral**, **Beneficial** or **Harmful**

1) Some mutations can have a **neutral effect** on a protein's **function**. They may have a neutral effect because:
   - The mutation changes a base in a triplet, but the **amino acid** that the triplet codes for **doesn't change**. This happens because **some amino acids** are coded for by **more than one triplet**. E.g. both **TAT** and **TAC** code for **tyrosine**, so if TAT is changed to TAC the **amino acid won't change**.
   - The mutation produces a triplet that codes for a **different amino acid**, but the amino acid is **chemically similar** to the original so it functions like the original amino acid. E.g. **arginine** (AGG) and **lysine** (AAG) are coded for by similar triplets — a **substitution** mutation can **swap** the amino acids. But this mutation would have a **neutral effect** on a **protein** as the amino acids are **chemically similar**.
   - The mutated triplet codes for an amino acid **not involved** with the protein's **function**, e.g. one that's located **far away** from an enzyme's **active site**, so the protein **works** as it **normally** does.
2) A **neutral effect** on protein function **won't** affect an **organism** overall.
3) However, some mutations **do** affect a protein's **function** — they can make a protein **more** or **less active**, e.g. by **changing** the **shape** of an enzyme's **active site**.
4) If protein function **is affected** it can have a **beneficial** or **harmful effect** on the **whole organism**:

Mutations with beneficial effects

- These have an **advantageous effect** on an organism, i.e. they **increase** its chance of **survival**.
- E.g. some bacterial **enzymes break down** certain **antibiotics**. **Mutations** in the genes that code for these enzymes could make them work on a **wider range** of antibiotics. This is **beneficial** to the **bacteria** because antibiotic resistance can help them to survive.

Mutations that are beneficial to the organism are passed on to future generations by the process of natural selection (see p. 64).

Mutations with harmful effects

- These have a **disadvantageous effect** on an organism, i.e. they **decrease** its chance of **survival**.
- E.g. **cystic fibrosis** (CF) can be caused by a **deletion** of three bases in the gene that codes for the **CFTR** (cystic fibrosis transmembrane conductance regulator) **protein**. The mutated CFTR protein **folds incorrectly**, so it's **broken down**. This leads to **excess mucus production**, which affects the **lungs** of CF sufferers.

Practice Questions

Q1 How does cAMP activate a protein?

Q2 Give two reasons why a mutation may have a neutral effect on a protein's function.

Amino acid	DNA triplet
Methionine	ATG
Tyrosine	TAT or TAC
Serine	TCA or TCC
Glycine	GGC or GGT
Cysteine	TGT or TGC

Exam Questions

Q1 a) Define the term mutation. [1 mark]
b) Describe two types of mutation that occur in DNA. [2 marks]

Q2 A gene begins with the following DNA sequence: ATGTATTCAGGCTGT
A mutation occurred where the ninth base was substituted by cytosine (C).
a) Write down the mutated DNA sequence. [1 mark]
b) Using the table, explain the effect that the mutation would have on the protein. [3 marks]

Mutations in adolescent turtles can enhance their ninja skills...

The important thing to remember about cAMP is that it alters the 3D structure of proteins — you can't just say that it activates proteins, you need to say how it does it as well. Don't forget that mutations can be harmless and that some can improve the way a protein functions — it's easy to associate 'mutation' with 'bad', but don't fall into that trap.

Meiosis

To kick-start this section we have a well-known A2-Biology crowd-pleaser... meiosis. If you thought mitosis was exciting at AS then you ain't seen nothing yet.

DNA from One Generation is Passed to the Next by Gametes

1) **Gametes** are the **sperm** cells in males and **egg** cells in females. They join together at **fertilisation** to form a **zygote**, which divides and develops into a **new organism**.
2) Normal **body cells** have the **diploid number** (**2n**) of chromosomes — meaning each cell contains **two** of each chromosome, one from the mum and one from the dad.
3) **Gametes** have the **haploid** (**n**) number — there's only one copy of each chromosome.
4) At **fertilisation**, a **haploid sperm** fuses with a **haploid egg**, making a cell with the **normal diploid number** of chromosomes (2n).

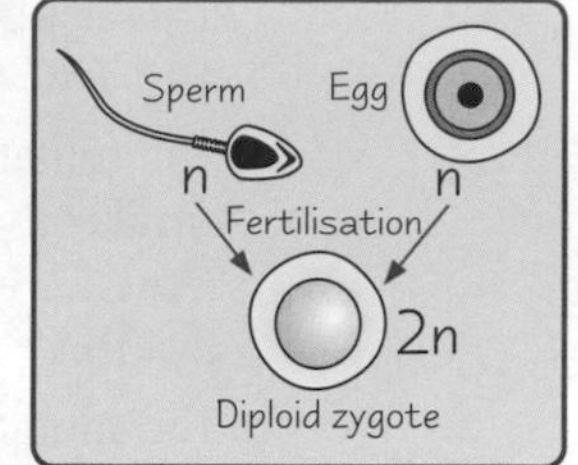

Gametes are Formed by Meiosis

Meiosis is a type of **cell division** that happens in the reproductive organs to **produce gametes**. Cells that divide by meiosis are **diploid** to start with, but the cells that are formed from meiosis are **haploid** — the chromosome number halves. Cells formed by meiosis are all **genetically different** because each new cell ends up with a **different combination** of chromosomes.

Gametes Divide Twice in Meiosis

Centromere
One chromatid
Sister chromatids

Before meiosis, **interphase** happens — the cell's DNA unravels and **replicates** so there are **two** copies of each chromosome in each cell. Each copy of the chromosome is called a **chromatid** and a pair are called **sister chromatids** — they're joined in the middle by a **centromere**. After interphase, the cells enter meiosis where they **divide twice** — the first division is called **meiosis I** and the second is called **meiosis II**. There are **four similar stages** to each division called **prophase**, **metaphase**, **anaphase** and **telophase**:

Meiosis I

Prophase I
The **chromosomes condense**, getting shorter and fatter. **Homologous chromosomes pair up** — number 1 with number 1, 2 with 2, 3 with 3 etc. **Crossing-over** occurs (see next page). Tiny bundles of protein called **centrioles** start moving to opposite ends of the cell, forming a network of protein fibres across it called the **spindle**. The **nuclear envelope** (the membrane around the nucleus) **breaks down**.

Metaphase I
The **homologous pairs line up** across the **centre** of the cell and **attach** to the **spindle fibres** by their **centromeres**.

Anaphase I
The **spindles contract**, pulling the **pairs apart** (**one** chromosome goes to **each end** of the cell).

Telophase I
A **nuclear envelope** forms around each group of chromosomes and the **cytoplasm divides** so there are now **two haploid daughter cells**.

Homologous pair
Nuclear envelope
Nucleus
Cell
Cytoplasm
2 × 2n
Prophase I
Centrioles
Metaphase I
Spindle fibres
Anaphase I
Telophase I
Chromosome number is halved
2 × n
2 × n

A pair of homologous chromosomes is also called a bivalent.

Meiosis II

The two daughter cells undergo **prophase II, metaphase II, anaphase II** and **telophase II** — which are pretty much the same as the ones in **meiosis I** except with **half** the number of chromosomes. In **anaphase II**, the **sister chromatids are separated** — each **new** daughter cell inherits **one chromatid** from **each chromosome**. **Four haploid** daughter cells are produced.

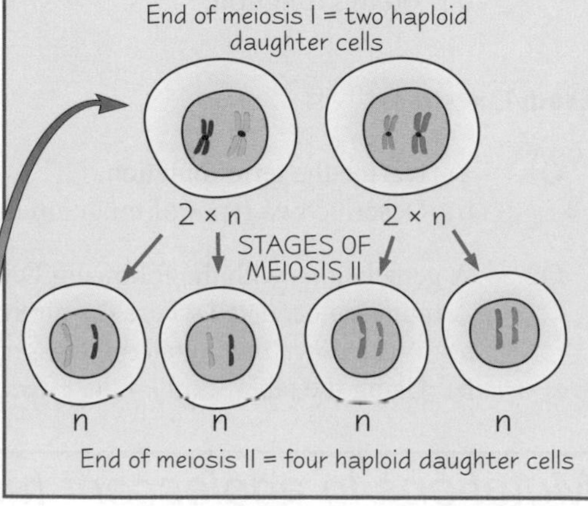

We've only shown 4 chromosomes — humans really have 46 (23 pairs).

Meiosis

Meiosis Produces Cells that are *Genetically Different*

Genetic variation is the **differences** that exist between **individuals' genetic material**. The reason meiosis is important is that it **creates** genetic variation — it makes gametes that are all genetically different. It does this in three ways:

1) Crossing-over of chromatids

1) During **prophase I**, **homologous chromosomes** come together and **pair up**.
2) In each pair, **one** chromosome is **maternal** (from your mum) and **one** is **paternal** (from your dad). They have the **same genes** but **different versions** of the genes, called **alleles**.
3) The **non-sister** chromatids twist around each other and **bits** of the **chromatids swap over** (they **break off** their chromatid and **join** onto the other chromatid).
4) The chromatids still contain the **same genes** but now have a **different combination** of **alleles**.
5) The **crossing-over** of chromatids during prophase I means that each of the **four daughter cells** formed from meiosis contains chromatids with a **different combination** of **alleles**.

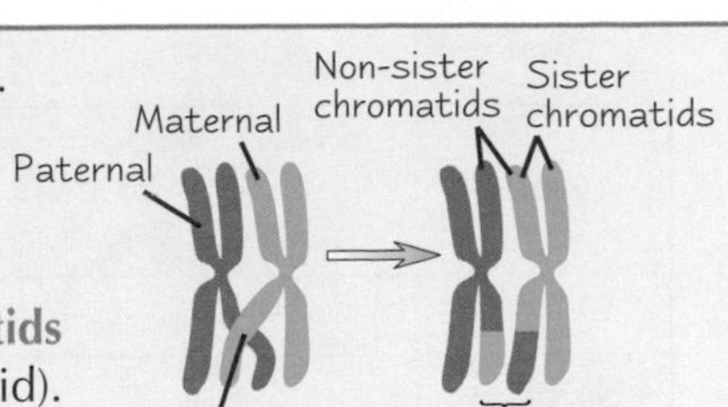

2) Independent assortment of chromosomes (in metaphase I)

1) During meiosis I, **different combinations** of maternal and paternal **chromosomes** go into each cell (e.g. one cell gets maternal chromosomes 1 and 2 and paternal 3, the other cell gets paternal 1 and 2, and maternal 3). So each cell ends up with a **different combination** of **alleles**.
2) If alleles are on the **same chromosome** they'll go into the same cell, so are **inherited together** — this is called **linkage**.

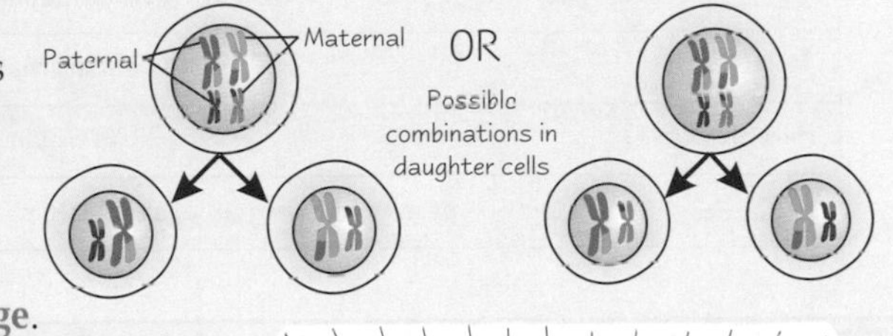

The way the chromosomes line up during metaphase I and the chromatids line up in metaphase II determines which chromosomes or chromatids will go into each cell.

3) Independent assortment of chromatids (in metaphase II)

During meiosis II, different **combinations** of **chromatids** go into each daughter cell. So each cell ends up with a **different combination** of **alleles**.

So, **crossing-over**, plus the **independent assortment of chromosomes** and **chromatids** during meiosis, means that **gametes** end up with a unique assortment of **alleles** (i.e. all the cells are **genetically different**). Then, during **fertilisation**, **any egg** can fuse with **any sperm**, which also **creates variation**. This means new individuals have a **new mixture** of alleles, making them **genetically unique**.

Practice Questions

Q1 Name the four stages of meiosis I.

Q2 At the end of meiosis II, are the daughter cells haploid or diploid?

Q3 At what stage of meiosis does crossing-over of chromatids occur?

Q4 What is linkage?

A

B

Exam Questions

Q1 a) Describe the behaviour of the chromosomes, nuclear envelope and centrioles in prophase I. [4 marks]
b) Name the stages of meiosis shown in the pictures above. [2 marks]

Q2 Humans show genetic variation due to meiosis and fertilisation.
a) Describe and explain three processes in meiosis that lead to genetic variation. [7 marks]
b) State how fertilisation increases genetic variation. [1 mark]

Physics — that's what I call crossing-over to the dark side...

You're probably sat there thinking about the good old days of AS, where meiosis didn't seem that hard... But, as your teachers will say, this is sooooo much more interesting. And I'm afraid that even if you don't agree with that, you still have to get your head around this lot. Go over it again and again until you start dreaming about chromosomes...

Inheritance

If you know the alleles two organisms have you can work out the alleles their offspring might have if they get jiggy with it.

You **Need to Know** *These* **Genetic Terms**

'Codes for' means 'contains the instructions for'.

TERM	DESCRIPTION
Gene	A sequence of bases on a DNA molecule that codes for a protein (polypeptide), which results in a characteristic, e.g. the gene for eye colour.
Allele	A different version of a gene. Most plants and animals, including humans, have two alleles of each gene, one from each parent. The order of bases in each allele is slightly different — they code for different versions of the same characteristic. They're represented using letters, e.g. the allele for brown eyes (B) and the allele for blue eyes (b).
Genotype	The alleles an organism has, e.g. BB, Bb or bb for eye colour.
Phenotype	The characteristics the alleles produce, e.g. brown eyes.
Dominant	An allele whose characteristic appears in the phenotype even when there's only one copy. Dominant alleles are shown by a capital letter. E.g. the allele for brown eyes (B) is dominant — if a person's genotype is Bb or BB, they'll have brown eyes.
Recessive	An allele whose characteristic only appears in the phenotype if two copies are present. Recessive alleles are shown by a lower case letter. E.g. the allele for blue eyes (b) is recessive — if a person's genotype is bb, they'll have blue eyes.
Codominant	Alleles that are both expressed in the phenotype — neither one is recessive, e.g. the alleles for haemoglobin.
Locus	The fixed position of a gene on a chromosome. Alleles of a gene are found at the same locus on each chromosome in a pair.
Homozygote	An organism that carries two copies of the same allele, e.g. BB or bb.
Heterozygote	An organism that carries two different alleles, e.g. Bb.
Carrier	A person carrying an allele which is not expressed in the phenotype but that can be passed on to offspring.

Genetic Diagrams Show *the* **Possible Genotypes** *of* **Offspring**

Individuals have **two alleles** for **each gene**. **Gametes** contain only **one allele** for each gene. When two gametes fuse together, the alleles they contain form the **offspring's genotype**. **Genetic diagrams** can be used to **predict** the **genotypes** and **phenotypes** of the offspring produced if two parents are **crossed** (bred). For example, the genetic diagram below shows how **wing length** is inherited in fruit flies. This is an example of **monohybrid inheritance** — the inheritance of a **single characteristic** (gene) controlled by **different alleles**.

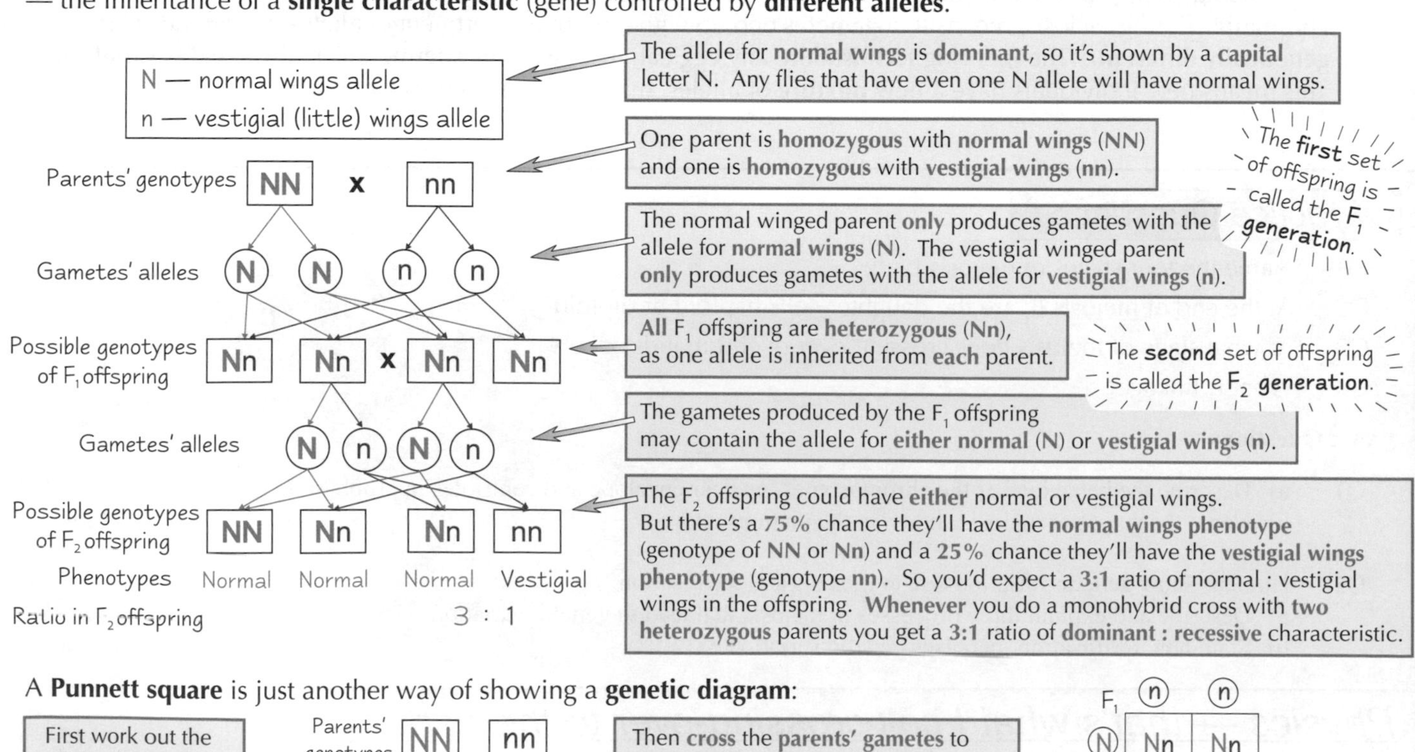

Inheritance

Some Genes Have Codominant Alleles

Occasionally, alleles show **codominance** — **both alleles** are expressed in the **phenotype**, **neither one** is recessive. One example in humans is the allele for **sickle-cell anaemia**:

1) People who are **homozygous** for **normal haemoglobin** (**H^NH^N**) don't have the disease.
2) People who are **homozygous** for **sickle haemoglobin** (**H^SH^S**) have **sickle-cell anaemia** — all their **blood cells** are **sickle-shaped** (crescent-shaped).
3) People who are **heterozygous** (**H^NH^S**) have an **in-between** phenotype, called the **sickle-cell trait** — they have **some** normal haemoglobin and some sickle haemoglobin. The two alleles are **codominant** because they're **both** expressed in the **phenotype**.
4) The **genetic diagram** shows the possible offspring from **crossing** two parents with **sickle-cell trait** (**heterozygous**).

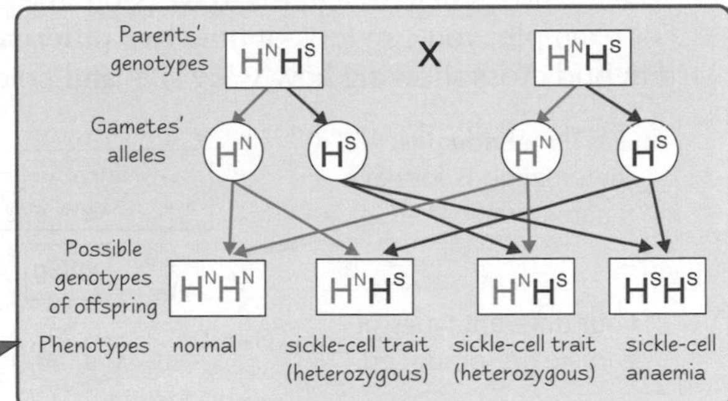

You need to be able to work out genetic diagrams for codominant alleles and sex-linked characteristics.

Some Characteristics are Sex-linked

1) The genetic information for **gender** (**sex**) is carried on two **sex chromosomes**.
2) In mammals, **females** have **two X** chromosomes (XX) and **males** have **one X** and **one Y** chromosome (XY).
3) A **characteristic** is said to be **sex-linked** when the allele that codes for it is located on a **sex chromosome**.
4) The **Y chromosome** is **smaller** than the X chromosome and carries **fewer genes**. So most genes on the sex chromosomes are **only carried** on the X chromosome (called **X-linked** genes).
5) As **males** only have **one X chromosome** they often only have **one allele** for sex-linked genes. So because they **only** have one copy they **express** the **characteristic** of this allele **even** if it's **recessive**. This makes males **more likely** than females to show **recessive phenotypes** for genes that are sex-linked.
6) Genetic disorders caused by **faulty alleles** on sex chromosomes include **colour blindness** and **haemophilia**. The faulty alleles for both of these disorders are carried on the X chromosome — they're called **X-linked disorders**.

Example **Colour blindness** is a **sex-linked disorder** caused by a faulty allele carried on the **X** chromosome. As it's sex-linked **both** the chromosome and the allele are **represented** in the **genetic diagram**, e.g. **X^n**, where **X** represents the **X chromosome** and **n** the **faulty allele** for **colour vision**. The **Y chromosome** doesn't have an allele for colour vision so is **just** represented by **Y**. **Females** would need **two copies** of the **recessive allele** to be colour blind, while **males** only need **one copy**. This means colour blindness is **much rarer** in **women** than **men**.

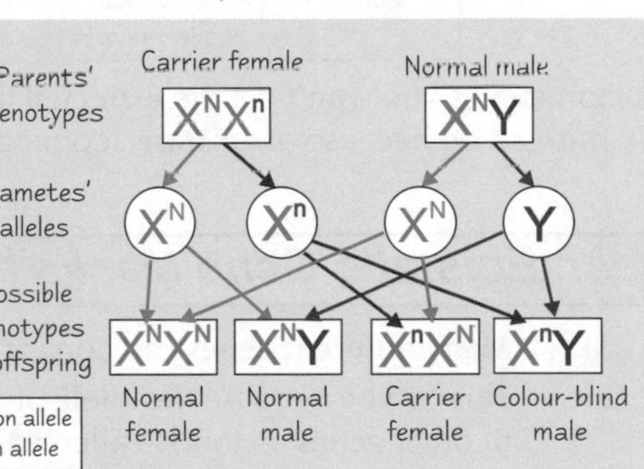

Practice Questions

Q1 What is meant by the terms genotype, phenotype, dominant allele and recessive allele?

Q2 What is a codominant allele?

Exam Question

Q1 Haemophilia A is a sex-linked genetic disorder caused by a recessive allele carried on the X chromosome (X^h).

a) Draw a genetic diagram for a female carrier and a male sufferer to predict their offspring's genotype. [3 marks]

b) Explain why haemophilia is more common in males than females. [3 marks]

If there's a dominant revision allele I'm definitely homozygous recessive...

OK, so there are a lot of fancy words on these pages and yes, you do need to know them all. Sorry about that. But don't despair — once you've learnt what the words mean and know how genetic diagrams work it'll all just fall into place.

Phenotypic Ratios and Epistasis

Right, this stuff is fairly hard, so if you don't get it first time don't panic.
Make sure you're happy with the genetic diagrams on the previous pages before you get stuck into these two.

Genetic Diagrams can Show how More Than One Characteristic is Inherited

You can use genetic diagrams to work out the chances of offspring inheriting certain **combinations** of characteristics. For example, you can look at how **two different genes** are inherited — **dihybrid inheritance**. The diagram below is a dihybrid cross showing how wing size **and** colour are inherited in **fruit flies**.

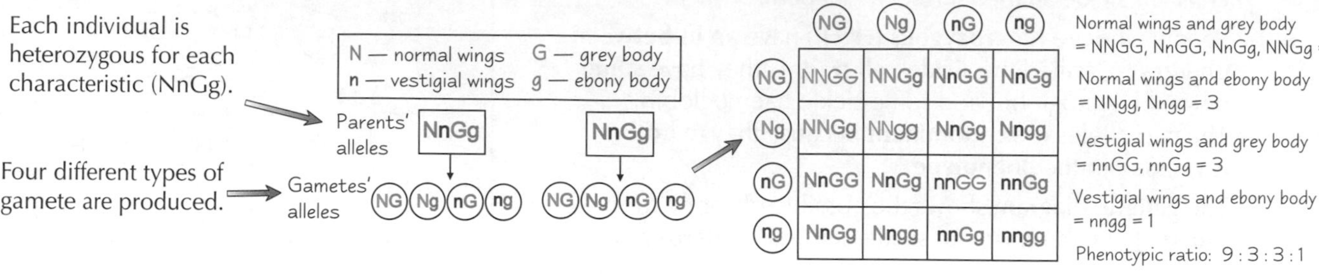

Phenotypic Ratios can be Predicted

The **phenotypic ratio** is the **ratio** of **different phenotypes** in offspring. Genetic diagrams allow you to **predict** the phenotypic ratios in **F_1 and F_2 offspring**. You need to **remember** the ratios for the following crosses:

Type of cross	Parents	Phenotypic ratio in F_1	Phenotypic ratio in F_2
Monohybrid	Homozygous dominant × homozygous recessive (e.g. NN × nn)	All heterozygous offspring (e.g. Nn)	3 : 1 dominant : recessive
Dihybrid	Homozygous dominant × homozygous recessive (e.g. NNGG × nngg)	All heterozygous offspring (e.g. NnGg)	9 : 3 : 3 : 1 dominant both : dominant 1st recessive 2nd : recessive 1st dominant 2nd : recessive both
Codominant	Homozygous for one allele × homozygous for the other allele (e.g. H^NH^N x H^SH^S)	All heterozygous offspring (e.g. H^NH^S)	1 : 2 : 1 homozygous for one allele : heterozygous : homozygous for the other allele

Sometimes you **won't** get the **expected** (predicted) phenotypic ratio — it'll be quite different. This can be because of **epistasis** (coming up next) or **linkage** (see page 55).

An Epistatic Gene Masks the Expression of Another Gene

1) **Many different genes** can control the **same** characteristic — they **interact** to form the phenotype.
2) This can be because the **allele** of one gene **masks** (blocks) **the expression** of the alleles of other genes — this is called **epistasis**.

I'm still dashing, even with my widow's peak.

Example 1 In humans a **widow's peak** (see picture) is controlled by one gene and **baldness** by others. If you have the **alleles** that code for baldness, it **doesn't matter** whether you have the allele for a widow's peak or not, as you have **no hair**. The baldness genes are **epistatic** to the widow's peak gene, as the baldness genes **mask** the expression of the widow's peak gene.

Example 2 **Flower pigment** in a plant is controlled by two genes. **Gene 1** codes for a **yellow pigment** (Y is the dominant yellow allele) and **gene 2** codes for an enzyme that **turns** the yellow pigment **orange** (R is the dominant orange allele). If you **don't have** the **Y** allele it **won't matter** if you have the R allele or not as the flower **will be colourless**. Gene 1 is **epistatic** to gene 2 as it can **mask** the expression of gene 2.

Colourless molecule → (gene 1 (YY or Yy)) → Yellow pigment → (gene 2 (RR or Rr)) → Orange pigment

3) **Crosses** involving epistatic genes **don't result** in the **expected phenotypic ratios** given above, e.g. if you cross **two heterozygous orange** flowered plants (YyRr) from the above example you wouldn't get the expected **9 : 3 : 3 : 1** phenotypic ratio for a **normal dihybrid cross** (see next page).

Phenotypic Ratios and Epistasis

You can Predict the Phenotypic Ratios for Some Epistatic Genes

Just as you can **predict** the phenotypic ratios for **normal dihybrid crosses** (see previous page), you can predict the phenotypic ratios for dihybrid crosses involving some **epistatic genes** too:

A dihybrid cross involving a recessive epistatic allele — 9 : 3 : 4

Having **two copies** of the **recessive** epistatic allele **masks** (**blocks**) the expression of the **other gene**. If you cross a **homozygous recessive** parent with a **homozygous dominant** parent you will get a **9 : 3 : 4** phenotypic ratio of **dominant both : dominant epistatic recessive other : recessive epistatic** in the **F_2 generation**.

E.g. the **flower example** from the **previous page** is an example of a **recessive epistatic allele**. If a plant is **homozygous recessive** for the **epistatic gene** (**yy**) then it will be **colourless**, **masking** the expression of the orange gene. So if you cross homozygous parents, you should get a **9 : 3 : 4** ratio of **orange : yellow : white** in the **F_2 generation**. You can check the **phenotypic ratio** is right **using a genetic diagram**:

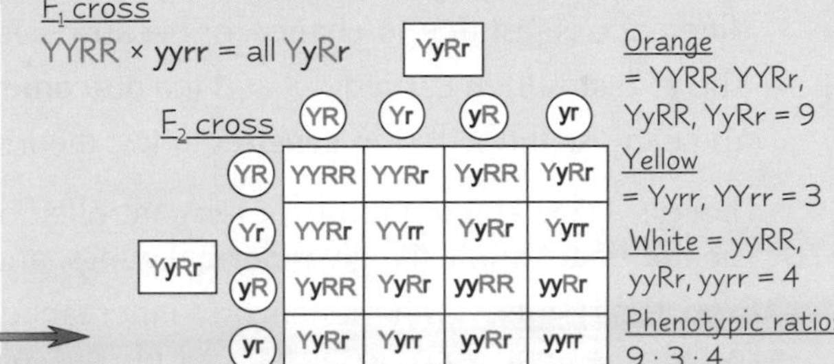

A dihybrid cross involving a dominant epistatic allele — 12 : 3 : 1

Having **at least one** copy of the **dominant epistatic** allele **masks** (**blocks**) the expression of the other gene. Crossing a **homozygous recessive** parent with a **homozygous dominant** parent will produce a **12 : 3 : 1** phenotypic ratio of **dominant epistatic : recessive epistatic dominant other : recessive both** in the F_2 generation.

E.g. **squash colour** is controlled by two genes — the **colour epistatic gene** (**W/w**) and the **yellow gene** (**Y/y**). The **no-colour, white** allele (**W**) is **dominant** over the **coloured** allele (**w**), so **WW** or **Ww** will be **white** and **ww** will be **coloured**. The yellow gene has the **dominant yellow** allele (**Y**) and the **recessive green** allele (**y**). So if the plant has **at least one W**, then the squash **will be white**, **masking** the expression of the yellow gene. So if you cross **wwyy** with **WWYY**, you'll get a **12 : 3 : 1** ratio of **white : yellow : green** in the F_2 generation. Here's a **genetic diagram** to prove it:

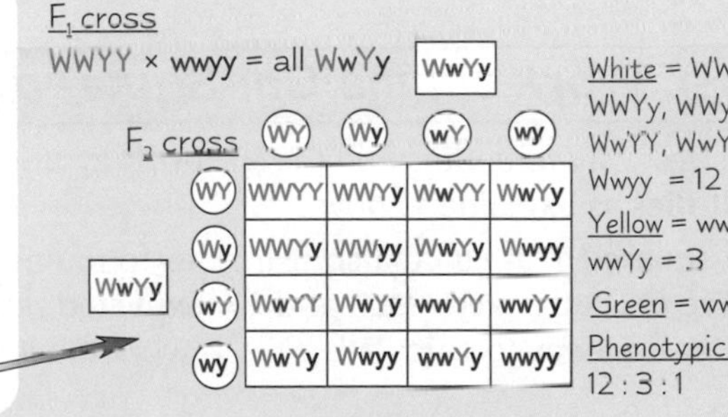

Practice Questions

Q1 What phenotypic ratio would be produced in the F_1 generation and the F_2 generation by the cross aabb × AABB (assuming no epistasis)?

Q2 Describe epistasis.

Exam Questions

Homozygous curly hair (hhss) crossed with a homozygous bald (HHSS)

Phenotypes of the F_2 offspring produced		
Bald	Straight hair	Curly hair
36	9	3

Q1 Colour (R red, r pink) and lines (G green, g white) are controlled by two genes in the Snozcumber plant. Draw a genetic diagram of the cross: homozygous for red and white lines × homozygous for pink and green lines. [3 marks]

Q2 Coat colour in mice is controlled by two genes. Gene 1 controls whether fur is coloured (C) or albino (c). Gene 2 controls whether the colour is grey (G) or black (g). Gene 1 is epistatic over gene 2. Describe and explain the phenotypic ratio produced in the F_2 generation from a CCGG × ccgg cross. [4 marks]

Q3 Hair type in Dillybopper beetles is controlled by two genes: hair (H bald, h hair) and type (S straight, s curly). The F_2 offspring of a cross are shown in the table. Explain the phenotypic ratio shown by the cross. [3 marks]

Biology students — 9 : 1 phenotypic ratio normal : geek...

I don't know about you but I think I need a lie-down after these pages. Epistasis is a bit of a tricky topic, but you just need to understand what it is and learn the phenotypic ratios for the different types of epistasis — dominant and recessive.

The Chi-Squared Test

Just when you thought it was safe to turn the page... I stick in some maths. Surprise!

The **Chi-Squared Test** Can Be Used to **Check** the **Results** of **Genetic Crosses**

1) The **chi-squared (χ^2) test** is a **statistical test** that's used to see if the **results** of an experiment **support** a **theory**.
2) First, the theory is used to **predict** a **result** — this is called the **expected result**.
   Then, the experiment is carried out and the **actual result** is recorded — this is called the **observed result**.
3) To see if the results support the theory you have to make a **hypothesis** called the **null hypothesis**.
4) The null hypothesis is always that there's **no significant difference** between the observed and expected results (your experimental result will usually be a bit different from what you expect, but you need to know if the difference is just **due to chance**, or because your **theory is wrong**).
5) The **χ^2 test** is then carried out and the **outcome** either **supports** or **rejects** the **null hypothesis**.
6) You can use the χ^2 test in **genetics** to test theories about the **inheritance** of **characteristics**. For example:

Theory: **Wing length** in fruit flies is controlled by a **single gene** with **two alleles** (**monohybrid inheritance**). The **dominant** allele (N) gives **normal** wings, and the **recessive** allele (n) gives **vestigial** wings.

Expected results: With monohybrid inheritance, if you cross a **homozygous dominant** parent with a **homozygous recessive** parent, you'd expect a **3 : 1 phenotypic ratio** of **normal : vestigial** wings in the F_2 generation (see p. 56).

Observed results: The **experiment** (of crossing a homozygous dominant parent with a homozygous recessive parent) is **carried out** on fruit flies and the **number of offspring** with normal and vestigial wings is **counted**.

Null hypothesis: There's **no significant difference** between the observed and expected results.

(If the χ^2 test shows the observed and expected results are **not significantly different** the null hypothesis is **accepted** — the data supports the **theory** that wing length is controlled by **monohybrid inheritance**.)

First, **Work** out the **Chi-Squared Value...**

The best way to understand the χ^2 test is to work through an example — here's one for testing the **wing length** of **fruit flies** as explained above.

Chi-squared χ^2 is calculated using this formula:

$$\chi^2 = \sum \frac{(O-E)^2}{E}$$

where **O** = **observed** result and **E** = **expected** result.

The easiest way to calculate χ^2 is to work it out in **stages** using a table:

You don't need to learn the formula for chi-squared — it'll be given to you in the exam.

1. First, the **number of offspring** (out of a total of 160) **expected** for each phenotype is worked out. E for normal wings: 160 (total) ÷ 4 (ratio total) × 3 (predicted ratio for normal wings) = 120. E for vestigial wings: 160 ÷ 4 × 1 = 40.

Phenotype	Ratio	Expected Result (E)	Observed Result (0)
Normal wings	3	120	
Vestigial wings	1	40	

2. Then the **actual number** of offspring **observed** with each phenotype (out of the 160 offspring) is **recorded**, e.g. 111 with normal wings.

Phenotype	Ratio	Expected Result (E)	Observed Result (0)
Normal wings	3	120	111
Vestigial wings	1	40	49

3. The results are used to work out χ^2, taking it **one step at a time**:

   a. First calculate **O – E** (**subtract** the **expected result** from the **observed result**) for each phenotype. E.g. for normal wings: 111 – 120 = –9.

   b. Then the resulting numbers are **squared**, e.g. $9^2 = 81$

   c. These figures are divided by the **expected results**, e.g. 81 ÷ 120 = 0.675.

   d. Finally, the numbers are **added** together to get χ^2, e.g. 0.675 + 2.025 = **2.7**.

Phenotype	Ratio	Expected Result (E)	Observed Result (0)	O – E	$(O-E)^2$	$\frac{(O-E)^2}{E}$
Normal wings	3	120	111	–9	81	0.675
Vestigial wings	1	40	49	9	81	2.025
					$\sum\frac{(O-E)^2}{E}$ =	2.7

Remember, you need to work it out for each phenotype first, then add all the numbers together.

The Chi-Squared Test

...Then Compare it to the Critical Value

1) To find out if there **is** no significant difference between your observed and expected results you need to **compare** your χ^2 **value** to a **critical value**.
2) The critical value is the value of χ^2 that corresponds to a 0.05 (**5%**) level of **probability** that the **difference** between the observed and expected results is **due to chance**.
3) If your χ^2 value is **smaller** than the critical value then there **is no significant difference** between the observed and expected results — the **null hypothesis** is **accepted**. E.g. for the example on the previous page the χ^2 value is **2.7**, which is **smaller** than the critical value of **3.84** — there's **no significant difference** between the observed and expected results. This means the **theory** that wing length in fruit flies is controlled by **monohybrid inheritance** is **supported**.
4) If your χ^2 value is **larger** than the critical value then there **is a significant difference** between the observed and expected results (something **other than chance** is causing the difference) — the **null hypothesis** is **rejected**.
5) In the exam you might be **given** the **critical value** or asked to **work it out** from a **table**:

Using a χ^2 table:

If you're not given the critical value, you may have to find it yourself from a χ^2 **table** — this shows a range of **probabilities** that correspond to different **critical values** for different **degrees of freedom** (explained below). Biologists normally use a **probability** level of **0.05** (5%), so you only need to look in that column.

- First, the **degrees of freedom** for the experiment are worked out — this is the **number of classes** (number of phenotypes) **minus one**. E.g. 2 – 1 = 1.
- Next, the **critical value** corresponding to a **probability** of **0.05** at **one degree of freedom** is found in the table — here it's **3.84**.
- Then just **compare** your χ^2 value of **2.7** to this critical value, as explained above.

degrees of freedom	no. of classes	Critical values					
1	2	0.46	1.64	2.71	3.84	6.64	10.83
2	3	1.39	3.22	4.61	5.99	9.21	13.82
3	4	2.37	4.64	6.25	7.82	11.34	16.27
4	5	3.36	5.99	7.78	9.49	13.28	18.47
probability that result is due to chance only		0.50 (50%)	0.20 (20%)	0.10 (10%)	0.05 (5%)	0.01 (1%)	0.001 (0.1%)

Practice Questions

Q1 What is a χ^2 test used for?

Q2 What can the results of the χ^2 test tell you?

Q3 How do you tell if the difference between your observed and expected results is due to chance?

Exam Question

Q1 A scientist is investigating petal colour in a flower. It's thought to be controlled by two separate genes (dihybrid inheritance), the colour gene — B = blue, b = purple, and the spots gene — W = white, w = yellow. A cross involving a homozygous dominant parent and a homozygous recessive parent should give a 9 : 3 : 3 : 1 ratio in the F_2 generation. The scientist observes the number of offspring showing each of four phenotypes in 240 F_2 offspring. Her results are shown in the table.

Her null hypothesis is that there is no significant difference between the observed and expected ratios.

a) Complete the table to calculate χ^2 for this experiment. [4 marks]

b) The critical value for this experiment is 7.82. Explain whether the χ^2 value supports or rejects the null hypothesis. [2 marks]

Phenotype	Ratio	Expected Result (E)	Observed Result (O)	O – E	O – E²	(O – E²) / E
Blue with white spots	9	135	131			
Purple with white spots	3	45	52			
Blue with yellow spots	3	45	48			
Purple with yellow spots	1	15	9			
					$\sum \frac{(O-E)^2}{E} =$	

The expected result of revising these pages — boredom...

...the observed result — boredom (except for the maths geeks among you). Don't worry if you're not brilliant at maths though, you don't have to be to do the chi-squared test — just make sure you know the steps above off by heart. You could even practise going through the example on these pages without looking at the book... go on, you know you want to.

Variation

Some people are tall, others are short. Some people wear glasses, others don't. Some people like peanut butter sandwiches, others... well, you get the picture. Basically variety is the spice of life and here's why we're all different.

Variation Exists Between All Individuals

1) **Variation** is the **differences** that exist between **individuals**. Every individual organism is **unique** — even **clones** (such as identical twins) show **some variation**.
2) Variation can occur **within species**, e.g. **individual** European robins weigh **between** 16 g and 22 g and show some variation in many other characteristics including length, wingspan, colour and beak size.
3) It can also occur **between species**, e.g. the **lightest** species of bird is the bee hummingbird, which weighs around 1.6 g on average and the **heaviest** species of bird is the ostrich, which can weigh up to 160 kg (100 000 times as much).

Variation — a concept lost on the army.

Variation can be Continuous...

1) **Continuous variation** is when the **individuals** in a population vary **within a range** — there are **no distinct categories**, e.g. **humans** can be **any height** within a range (139 cm, 175 cm, 185.9 cm, etc.), not just tall or short.
2) Some more examples of continuous variation include:
   - **Finger length** — e.g. a human finger can be any length within a range.
   - **Plant mass** — e.g. the mass of the seeds from a flower head varies within a range.

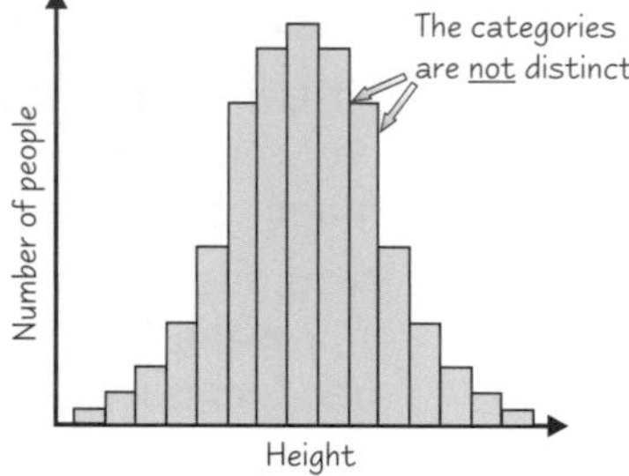

...or Discontinuous

1) **Discontinuous variation** is when there are two or more **distinct categories** — each individual falls into **only one** of these categories, there are **no intermediates**.
2) Here are some examples of discontinuous variation:
   - **Sex** — e.g. animals can be either male or female.
   - **Blood group** — e.g. humans can be group A, B, AB or O.

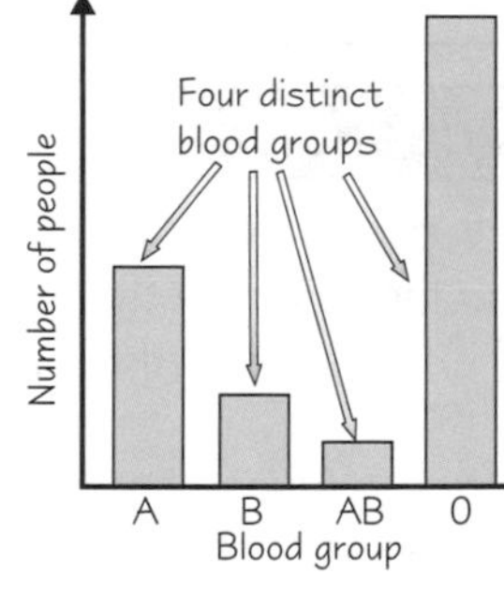

Variation can be Influenced by Your Genes...

1) **Different species** have **different genes**.
2) Individuals of the **same species** have the **same genes**, but **different versions** of them (called **alleles**).
3) The genes and alleles an organism has make up its **genotype**.
4) The **differences** in **genotype** result in **variation** in **phenotype** — the **characteristics** displayed by an organism. (Variation in phenotype is also referred to as **phenotypic variation**.)

EXAMPLE	**Human blood group** — there are **three** different **blood group alleles**, which result in **four different blood groups**.

5) **Inherited** characteristics that show **continuous** variation are usually **influenced** by **many genes** — these characteristics are said to be **polygenic**. For example, **human skin colour** is polygenic — it comes in **loads** of **different shades** of colour.
6) **Inherited** characteristics that show **discontinuous** variation are usually influenced by only **one gene** (or a **small number** of genes), e.g. **violet flower colour** (either coloured or white) is controlled by only one gene. Characteristics controlled by **only one gene** are said to be **monogenic**.

Variation

...the Environment...

Variation can also be caused by **differences in the environment**, e.g. climate, food, lifestyle. Characteristics controlled by environmental factors can **change** over an organism's life.

EXAMPLES

1) **Accent** — this is determined by **environmental factors only**, including **where you live** now, where you **grew up** and the accents of **people around you.**
2) **Pierced ears** — this is also **only** determined by **environmental factors**, e.g. **fashion, peer pressure.**

...or Both

Genetic factors determine genotype and the characteristics an organism's **born with**, but **environmental factors** can **influence** how some characteristics **develop**. Most phenotypic variation is caused by the **combination** of **genotype** and **environmental factors**. Phenotypic variation influenced by both usually shows **continuous variation**.

EXAMPLES

1) **Height of pea plants** — pea plants come in **tall** and **dwarf** forms (**discontinuous** variation), which is determined by **genotype**. However, the **exact height** of the tall and dwarf plants **varies** (**continuous** variation) because of **environmental factors** (e.g. **light intensity** and **water availability** affect how tall a plant grows).

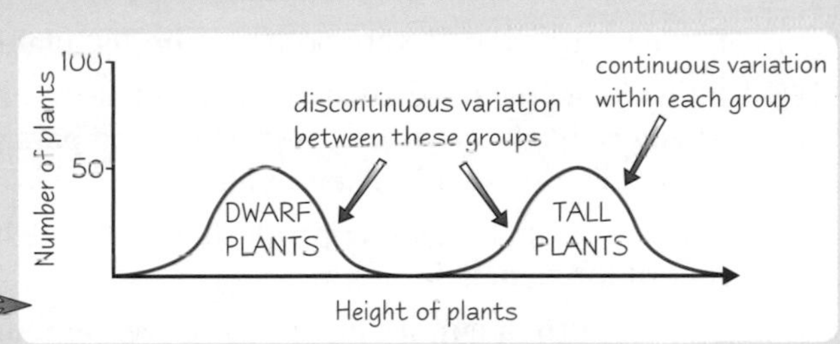

2) **Human body mass** — this is **partly genetic** (large parents often have large children), but it's also **strongly influenced** by **environmental factors**, like **diet** and **exercise**. Body mass **varies** within a **range**, so it's **continuous** variation.

Practice Questions

Q1 What is variation?

Q2 Describe what is meant by discontinuous variation and give one example.

Q3 Briefly describe what is meant by variation caused by genotype.

Puppy	Mass / kg	Colour	Puppy	Mass / kg	Colour	Puppy	Mass / kg	Colour
1	10.04	yellow	6	10.39	yellow	11	9.25	black
2	10.23	chocolate	7	10.55	chocolate	12	11.06	black
3	15.65	black	8	15.87	chocolate	13	12.45	yellow
4	18.99	black	9	16.99	black	14	14.99	yellow
5	9.45	black	10	10.47	yellow	15	10.93	chocolate

Exam Questions

Q1 The mass and coat colour of 15 Labrador puppies is shown in the table.

a) What type of variation (continuous or discontinuous) is shown by the coat colour of the puppies? [1 mark]

b) Calculate the range of puppy mass. [1 mark]

c) Which of the characteristics described in the table is most likely to be influenced by both genotype and the environment? Explain your answer. [2 marks]

Q2 Give an example of a human characteristic influenced by both genotype and the environment. Explain your answer. [2 marks]

Revision boredom shows discontinuous variation — always bored with it...

Hopefully you remember a lot of the info on these pages from AS, but I'm afraid you still need to know it off by heart for your A2 exam. Test yourself on examples of continuous and discontinuous variation — you never know when a sneaky question could pop up on them. Then, rest your brain so it's well and truly ready for a bit of evolution...

Evolution by Natural Selection and Genetic Drift

Variation between individuals of a species means that some organisms are better adapted to their environment than others — so they're more likely to survive and reproduce. Which leads us nicely on to evolution by natural selection...

Evolution is a Change in Allele Frequency

1) The complete range of **alleles** present in a **population** is called the **gene pool**.
2) **New alleles** are usually generated by **mutations** in **genes**.
3) How **often** an **allele occurs** in a population is called the **allele frequency**. It's usually given as a **percentage** of the total population, e.g. 35%, or a **number**, e.g. 0.35.
4) The **frequency** of an **allele** in a population **changes** over time — this is **evolution**.

A population is a group of organisms of the same species living in a particular area.

Evolution Occurs by Natural Selection

Variation is generated by meiosis and mutations.

1) **Individuals** within a population **vary** because they have **different alleles**.
2) **Predation**, **disease** and **competition** (**selection pressures**) create a **struggle for survival**.
3) Because individuals vary, some are **better adapted** to the selection pressures than others.
4) Individuals that have an allele that **increases** their **chance of survival** (a **beneficial** allele) are **more likely** to **survive**, **reproduce** and **pass on** the beneficial allele, than individuals with different alleles.
5) This means that a **greater proportion** of the next generation **inherit** the **beneficial allele**.
6) They, in turn, are **more likely** to **survive**, **reproduce** and **pass on** their genes.
7) So the **frequency** of the beneficial allele **increases** from generation to generation.
8) This process is called **natural selection**.

A selection pressure is anything that affects an organism's chance of survival and reproduction.

The *Environment* Affects *Which Characteristics Become More Common*

Whether the **environment** is **changing** or **stable** affects **which characteristics are selected for** by natural selection:

When the **environment isn't changing** much, individuals with alleles for characteristics towards the **middle** of the range are more likely to **survive** and **reproduce**. This is called **stabilising selection** and it **reduces the range** of possible **phenotypes**.

<u>EXAMPLE</u> In any **mammal population** there's a **range** of **fur length**. In a **stable climate**, having fur at the **extremes** of this range **reduces** the **chances** of **surviving** as it's harder to maintain the **right body temperature**. Animals with alleles for **average fur length** are the **most** likely to **survive**, **reproduce** and **pass on** their alleles. So these alleles **increase** in **frequency**. The **proportion** of the **population** with **average fur length increases** and the **range** of fur lengths **decreases**.

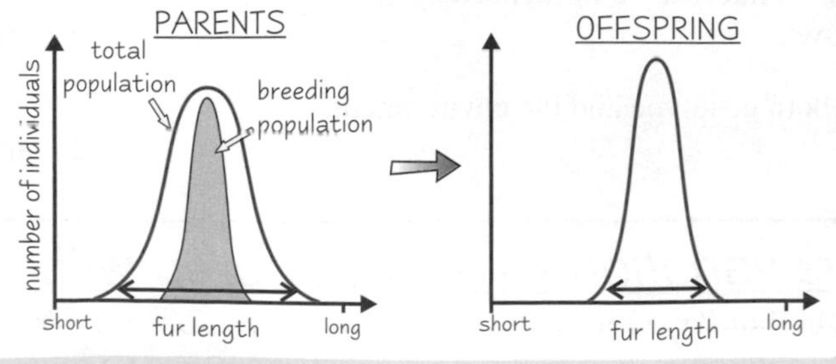

When there's a **change** in the environment, individuals with alleles for characteristics of an **extreme type** are more likely to **survive** and **reproduce**. This is called **directional selection**.

<u>EXAMPLE</u> If the environment becomes **very cold**, individual mammals with **alleles** for **long fur length** will find it **easier** to **maintain** the **right body temperature** than animals with short fur length. So they're **more likely** to **survive**, **reproduce** and **pass on** their alleles. Over time the **frequency** of alleles for **long fur length increases**.

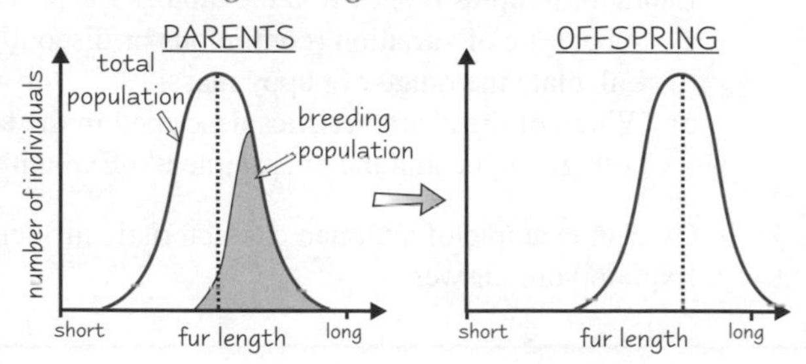

Evolution by Natural Selection and Genetic Drift

Evolution Also Occurs via Genetic Drift

1) **Natural selection** is just **one** process by which **evolution** occurs.
2) Evolution **also** occurs due to **genetic drift** — instead of **environmental factors** affecting which individuals **survive**, **breed** and pass on their alleles, **chance** dictates **which alleles** are **passed on**. Here's how it works:

- Individuals within a population show **variation** in their **genotypes** (e.g. A and B). ⟹
- By **chance**, the **allele** for **one genotype** (B) is **passed on** to the offspring **more often** than others.
- So the number of individuals with the allele **increases**. ⟹
- If by chance the same allele is passed on more often again and again, it can lead to **evolution** as the allele becomes **more common** in the population. ⟹

genotype A (4)
genotype B (4)

genotype A (3)
genotype B (5)

genotype A (1)
genotype B (7)

3) Natural selection and genetic drift work **alongside each other** to drive evolution, but one process can drive evolution **more** than the other depending on the **population size**.
4) **Evolution by genetic drift** usually has a **greater effect** in **smaller populations** where **chance** has a **greater influence**. In larger populations any chance factors tend to **even out** across the whole population.
5) The evolution of **human blood groups** is a good example of **genetic drift**:

Different **Native American tribes** show different **blood group frequencies**. For example, **Blackfoot Indians** are mainly **group A**, but **Navajos** are mainly **group O**. Blood group doesn't affect **survival** or **reproduction**, so the differences **aren't** due to evolution by natural selection. In the past, human populations were much **smaller** and were often found in **isolated groups**. The blood group differences were due to evolution by genetic drift — by **chance** the allele for **blood group O** was **passed on more often** in the **Navajo tribe**, so over time this **allele** and blood group became **more common**.

6) Evolution by genetic drift also has a greater effect if there's a **genetic bottleneck** — e.g. when a large population **suddenly becomes smaller** because of a **natural disaster**. For example:

The **mice** in a **large population** are either **black or grey**. The coat colour **doesn't** affect their **survival** or **reproduction**. A **large flood** hits the population and the **only survivors** are **grey** mice and **one black** mouse. **Grey** becomes the **most common colour** due to **genetic drift**.

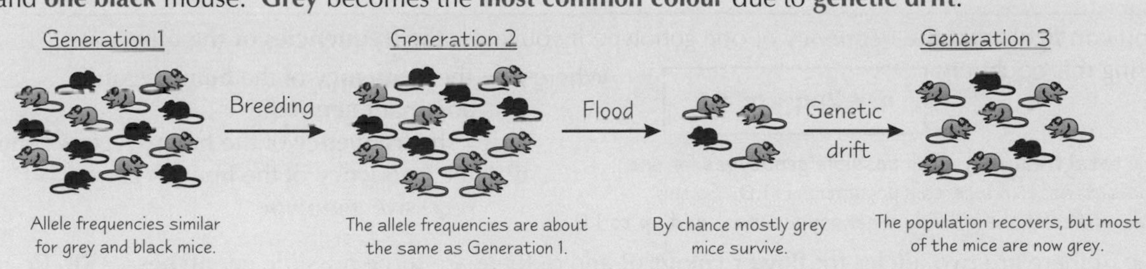

Practice Questions

Q1 What is evolution?

Q2 What is allele frequency?

Q3 What is genetic drift?

Exam Question

Q1 Before the 1800s, there were more pale-coloured peppered moths than dark peppered moths in Manchester. The pale moths were camouflaged on the trees they lived on. During the 1800s, air pollution in Manchester rose and blackened many of the trees. By the end of the 1800s, dark moths had become more common. Explain how natural selection gave rise to the increase in dark moths. [5 marks]

I've evolved to revise for hours and still not remember things...

The trickiest thing here is tying all the information together in your head. Basically, natural selection and genetic drift drive evolution. And the characteristics selected for in natural selection are determined by what the environment's like.

Hardy-Weinberg Principle and Artificial Selection

Now you know what allele frequency is you need to be able to calculate it. So switch your maths brain on now. Then you can take a breather on the right-hand page and learn all about artificial selection.

The **Hardy-Weinberg Principle** Predicts That **Allele Frequencies Won't Change**

1) The **Hardy-Weinberg principle** predicts that the **frequencies** of **alleles** in a population **won't change** from **one generation** to the **next**.
2) But this prediction is **only true** under **certain conditions** — it has to be a **large population** where there's **no immigration**, **emigration**, **mutations** or **natural selection**. There also needs to be **random mating** — all possible genotypes can breed with all others.
3) The **Hardy-Weinberg equations** (see below) are based on this principle. They can be used to **estimate the frequency** of particular **alleles** and **genotypes** within populations.
4) If the allele frequencies **do change** between generations in a large population then immigration, emigration, natural selection or mutations have happened.

The **Hardy-Weinberg Equations** Can be Used to **Predict Allele Frequency**...

1) You can **figure out** the frequency of one allele if you **know the frequency of the other**, using this equation:

$$p + q = 1$$

Where: **p** = the **frequency** of the **dominant** allele
q = the **frequency** of the **recessive** allele

The **total frequency** of **all possible alleles** for a characteristic in a certain population is **1.0**. So the frequencies of the **individual alleles** (the dominant one and the recessive one) must **add up to 1.0**.

2) E.g. a species of plant has either **red** or **white** flowers. Allele **R** (red) is **dominant** and allele **r** (white) is **recessive**. If the frequency of **R** is **0.4**, then the frequency of **r** is 1 – 0.4 = **0.6**.

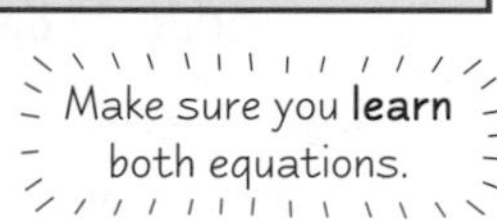

...**Genotype Frequency**...

1) You can **figure out** the frequency of one genotype if you **know the frequencies of the others**, using this equation:

$$p^2 + 2pq + q^2 = 1$$

Where p^2 = the **frequency** of the **homozygous dominant genotype**
$2pq$ = the **frequency** of the **heterozygous genotype**
q^2 = the **frequency** of the **homozygous recessive genotype**

The **total frequency** of **all possible genotypes** for one characteristic in a certain population is **1.0**. So the frequencies of the **individual genotypes** must **add up to 1.0**.

2) E.g. if there are **two alleles** for **flower colour** (R and r), there are **three possible genotypes** — **RR**, **Rr** and **rr**. If the frequency of genotype **RR** (p^2) is **0.34** and the frequency of genotype **Rr** (**2pq**) is **0.27**, the frequency of genotype **rr** (q^2) must be 1 – 0.34 – 0.27 = **0.39**.

...and the **Percentage** of a **Population** that has a **Certain Genotype**

EXAMPLE

The **frequency** of **cystic fibrosis** (genotype ff) in the UK is currently approximately **1 birth in 2000**. From this information you can estimate the **proportion** of people in the UK that are cystic fibrosis **carriers (Ff)**. To do this you need to find the **frequency** of **heterozygous genotype Ff**, i.e. **2pq**, using **both** equations:

First calculate q:
- Frequency of cystic fibrosis (homozygous recessive, ff) is 1 in 2000
- ff = $q^2 = 1 \div 2000 = 0.0005$
- So, $q = \sqrt{0.0005} = 0.022$

Next calculate p:
- using $p + q = 1$, $p = 1 - q$
- $p = 1 - 0.022$
- $p = 0.978$

Then calculate 2pq:
- $2pq = 2 \times 0.978 \times 0.022$
- $2pq = 0.043$

The **frequency** of **genotype Ff** is **0.043**, so the **percentage** of the UK population that are **carriers** is **4.3%**.

Hardy-Weinberg Principle and Artificial Selection

Artificial Selection Involves Breeding Individuals with Desirable Traits

Artificial selection is when **humans select individuals** in a population to **breed together** to get **desirable traits**. There are two examples you need to **learn**:

Artificial selection is also called selective breeding, which you might remember from AS.

Modern Dairy Cattle

Modern **dairy cows** produce **many litres of milk** a day as a result of **artificial selection**:

1) Farmers **select** a **female** with a **very high milk yield** and a **male** whose **mother** had a very high milk yield and **breed** these two **together**.
2) Then they **select** the **offspring** with the **highest milk yields** and **breed** them **together**.
3) This is continued over **several generations** until a **very high milk-yielding cow** is produced.

Bread Wheat

Bread wheat (*Triticum aestivum*) is the plant from which **flour** is produced for **bread-making**. It produces a **high yield** of wheat because of **artificial selection** by **humans**:

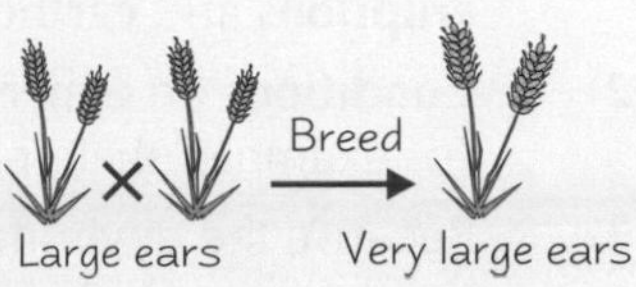

1) Wheat plants with a **high wheat yield** (e.g. large ears) are **bred together**.
2) The **offspring** with the **highest yields** are then **bred together**.
3) This is continued over **several generations** to produce a plant that has a **very high yield**.

You Need to be Able to Compare Natural Selection and Artificial Selection

You need to be able to describe the **similarities** and **differences** between **natural** and **artificial selection**:

Similarities:

- Both change the **allele frequencies** in the next generation — the **alleles** that **code** for the **beneficial/desirable characteristic** will become **more common** in the next generation.
- Both may make use of **random mutations** when they occur — if a random mutation produces an **allele** that gives a **beneficial/desirable phenotype**, it will be **selected for** in the next generation.

Differences:

- In natural selection, the organisms that reproduce are **selected by the environment** but in artificial selection this is **carried out by humans**.
- Artificial selection aims for a **predetermined result**, e.g. a farmer aims for a higher yield of milk, but in natural selection the **result** is **unpredictable**.
- Natural selection makes the species **better adapted** to the **environment**, but artificial selection makes the species **more useful** for **humans**.

Practice Questions

Q1 Which term represents the frequency of the dominant allele in the Hardy-Weinberg equations?

Q2 Describe two similarities between natural and artificial selection.

Exam Questions

Q1 A species of dog has either a black or brown coat. Allele B (black) is dominant and allele b (brown) is recessive. If the frequency of the b allele is 0.23, what is the frequency of the B allele? [1 mark]

Q2 Modern beef cattle (raised for meat production) produce a very high meat yield. Explain how artificial selection by farmers could have led to this. [3 marks]

This stuff's surely not that bad — Hardly worth Weining about...

Not many of you will be thrilled with the maths content on the left-hand page, but don't worry 'cause you just need to know the equations off by heart and what the terms in them mean. Then in the exam you'll be able to put the numbers in the correct places in the equation and, hey presto, you'll have your answer. Oh, and don't forget to take a calculator...

Speciation

Evolution leads to the development of lots of different species. Unfortunately for some species, the biologists had run out of good names, e.g. Colon rectum *(a type of beetle) and* Aha ha *(an Australian wasp). Oh dear.*

Speciation is the Development of a New Species

1) A **species** is defined as a group of **similar organisms** that can **reproduce** to give **fertile offspring**.
2) **Speciation** is the development of a **new species**.
3) It occurs when **populations** of the **same species** become **reproductively isolated** — changes in allele frequencies cause changes in phenotype that mean they can **no longer breed** together to produce **fertile offspring**.

Geographical Isolation and Natural Selection Lead to Speciation

1) Geographical isolation happens when a **physical barrier divides** a population of a species — **floods**, **volcanic eruptions** and **earthquakes** can all cause barriers that isolate some individuals from the main population.
2) **Conditions** on either side of the barrier will be slightly **different**. For example, there might be a **different climate** on each side.
3) Because the environment is different on each side, **different characteristics** will become **more common** due to **natural selection** (because there are **different selection pressures**):
   - Because different **characteristics** will be **advantageous** on each side, the **allele frequencies** will change in each population, e.g. if one allele is more advantageous on one side of the barrier, the frequency of that allele on that side will **increase**.
   - **Mutations** will take place **independently** in each population, also changing the **allele frequencies**.
   - The changes in allele frequencies will lead to changes in **phenotype frequencies**, e.g. the advantageous characteristics (**phenotypes**) will become more common on that side.
4) Eventually, individuals from different populations will have changed so much that they won't be able to breed with one another to produce **fertile** offspring — they'll have become **reproductively isolated**.
5) The two groups will have become separate **species**.

Geographical isolation is also known as ecological isolation.

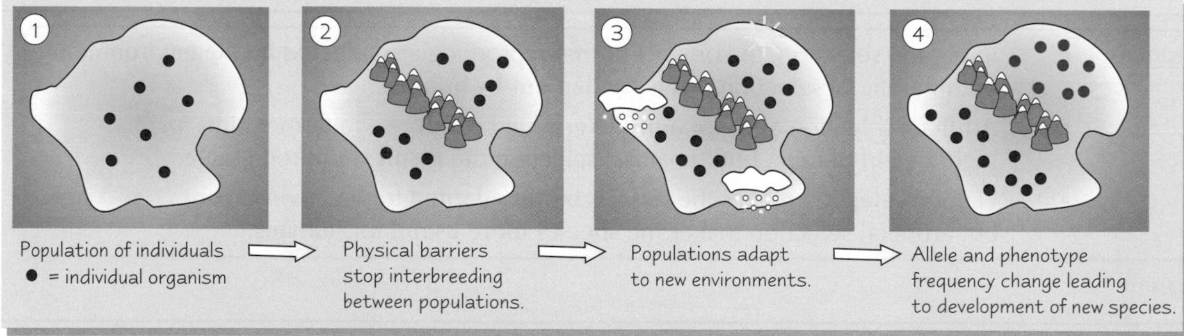

Reproductive Isolation Occurs in Many Ways

Reproductive isolation occurs because the **changes** in the alleles and phenotypes of the two populations **prevent** them from **successfully breeding together**. These changes include:

1) **Seasonal changes** — individuals from the same population develop different **flowering** or **mating** seasons, or become **sexually active** at **different times** of the year.
2) **Mechanical changes** — changes in **genitalia** prevent successful mating.
3) **Behavioural changes** — a group of individuals develop **courtship rituals** that **aren't attractive** to the main population.

A population **doesn't** have to become **geographically isolated** to become **reproductively isolated**. Random mutations could occur **within a population**, resulting in the changes mentioned above, **preventing** members of that population breeding with other members of the species.

Janice's courtship ritual was still successful in attracting mates.

Speciation

There are *Different Ways* to *Classify Species*

1) The traditional definition of a species is a group of **similar organisms** that can **reproduce** to give **fertile offspring**. This way of defining a species is called the **biological species concept**.
2) Scientists can have problems when using this definition, e.g. problems deciding **which species** an organism belongs to or if it's a new, **distinct species**.
3) This is because you can't always see their **reproductive behaviour** — you can't always tell if different organisms can reproduce to give **fertile offspring**. For example:

    1) They might be **extinct**, so you **can't** study their reproductive behaviour.
    2) They might **reproduce asexually** — they never **reproduce together** even if they belong to the same species, e.g. bacteria.
    3) There might be **practical** and **ethical issues** involved — you can't see if some organisms reproduce successfully in the wild (due to geography) and you can't study them in a lab (because it's unethical, e.g. humans and chimps are classed as separate species but has anyone ever tried mating them...).

4) Because of these problems, scientists sometimes use the **phylogenetic species concept** to classify organisms.
5) Phylogenetics is the **study** of the **evolutionary history** of groups of organisms (you might remember it from AS).
6) All organisms have **evolved** from shared common ancestors (**relatives**). The **more closely related** two species are, the **more recently** their last common ancestor will be.
7) Phylogenetics tells us **what's related** to what and how **closely related** they are.
8) Scientists can use phylogenetics to decide **which species** an organism belongs to or if it's a **new species** — if it's **closely related** to members of another species then it's probably the **same species**, but if it's **quite different** to any known species it's probably a **new species**.
9) There are also **problems** with classifying organisms using this concept, e.g. there's no cut-off to say how different two organisms have to be to be different species. For example, **chimpanzees** and **humans** are **different species** but about **94%** of our DNA is exactly the **same**.

The phylogenetic concept is also called the cladistic or evolutionary species concept.

Practice Questions

Q1 What is speciation?

Q2 What two concepts can be used to classify a species?

Exam Question

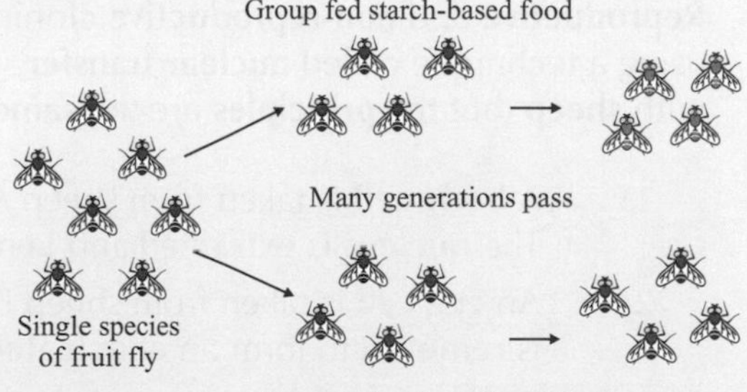

Q1 The diagram shows an experiment conducted with fruit flies. One population was split in two and each population was fed a different food. After many generations the two populations were placed together and it was observed that they were unable to breed together.

a) What evidence shows that speciation occurred? [1 mark]

b) Explain why the experiment resulted in speciation. [3 marks]

c) Suggest two possible reasons why members of the two populations were not able to breed together. [2 marks]

d) During the experiment, populations of fruit flies were artificially isolated. Suggest one way that populations of organisms could become isolated naturally. [1 mark]

Chess club members — self-enforced reproductive isolation...

These gags get better and better... Anyway, it's a bit of a toughie getting your head round the different mechanisms that can produce a new species. It doesn't help that reproductive isolation can happen on its own OR as a result of geographical isolation. Also, when reproductive isolation is caused by seasonal changes it's sometimes called seasonal isolation.

Cloning

Please don't try doing this at home — you'll only confuse people if there are 27 copies of you in the house...

Cloning makes Cells or Organisms Genetically Identical to Another Organism

Cloning is the process of producing **genetically identical cells** or **organisms** from the cells of an **existing organism**. Cloning can occur **naturally** in some **plants** and **animals**, but it can also be carried out **artificially**. You need to know about the **two types** of **artificial cloning** used for **animals**:

Reproductive cloning

1) **Reproductive cloning** is used to make a **complete organism** that's **genetically identical** to **another organism**.
2) Scientists use cloned animals for **research purposes**, e.g. they can **test new drugs** on cloned animals. They're all genetically identical, so the **variables** that come from **genetic differences** (e.g. the likelihood of developing cancer) are **removed**. This means the **results** are more **reliable**.
3) Reproductive cloning can be used to **save endangered animals** from **extinction** by cloning new individuals.
4) It can also be used by **farmers** to **increase** the **number** of animals with **desirable characteristics** to **breed from**, e.g. a prize-winning cow with high milk production could be cloned.
5) Loads of different animals have been cloned, e.g. **sheep**, **cattle**, **pigs** and **horses**.

Non-reproductive cloning

1) **Non-reproductive cloning** is used to make **embryonic stem cells** that are **genetically identical** to **another organism**. It's also called **therapeutic cloning**.
2) Embryonic stem cells are harvested from young **embryos**.
3) They have the **potential** to become **any cell type** in an organism, so scientists think they could be used to **replace damaged tissues** in a **range** of **diseases**, e.g. heart disease, spinal cord injuries, degenerative brain disorders like Parkinson's disease.
4) If replacement tissue is made from cloned embryonic stem cells that are **genetically identical** to the **patient's own cells** then the tissue **won't be rejected** by their immune system.

Take a look back at the stuff you learnt about stem cells at AS.

Animals are Artificially Cloned by Nuclear Transfer

Reproductive and **non-reproductive** cloning are **both carried out** using a technique called **nuclear transfer**. Here's how it's done with **sheep** (but the **principles** are the **same** for **any animal**):

1) A **body cell** is taken from sheep A. The **nucleus** is **extracted** and **kept**.
2) An **egg cell** is taken from sheep B. Its nucleus is **removed** to form an **enucleated egg cell**.
3) The nucleus from sheep A is **inserted** into the enucleated egg cell — the egg cell from **sheep B** now contains the **genetic information** from **sheep A**.
4) The egg cell is **stimulated** to **divide** and an **embryo** is formed.
5) In **reproductive cloning** the embryo is **implanted** into a **surrogate mother**. A **lamb** is produced that's a **genetically identical** copy of **sheep A**.
6) In **non-reproductive cloning stem cells** are **harvested** from the embryo. The stem cells are **genetically identical** to the cells in **sheep A**.

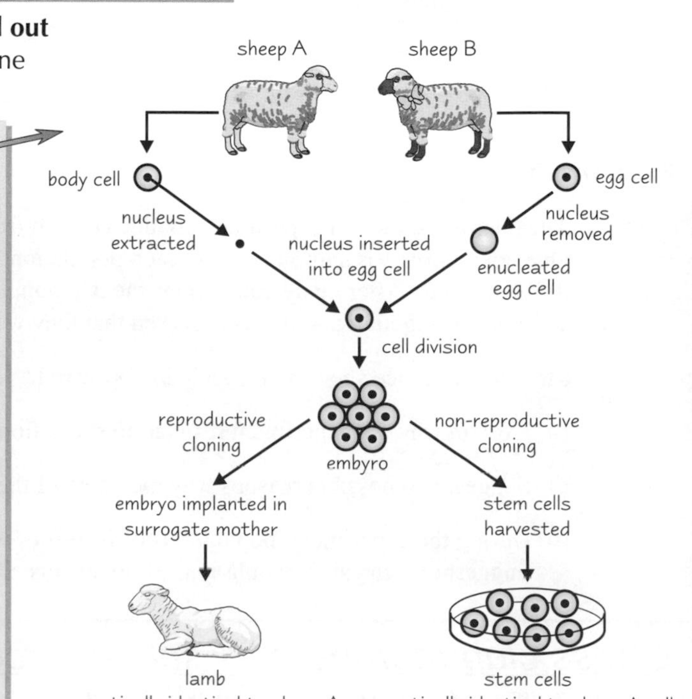

Cloning

Cloning Animals has Advantages and Disadvantages

Advantages

- **Desirable genetic characteristics** (e.g. high milk production in cows) are **always passed on** to clones — this **doesn't always** happen with **sexual reproduction**.
- **Infertile animals** can be **reproduced**.
- **Animals** can be **cloned** at **any time** — farmers wouldn't have to wait until a breeding season to produce new animals.

Disadvantages

- **Undesirable genetic characteristics** (e.g. a weak immune system) are **always passed on** to clones.
- Reproductive cloning is very **difficult**, **time-consuming** and **expensive** — **Dolly the sheep** was created after **277** nuclear transfer **attempts**.
- Some evidence suggests that clones **may not live as long** as natural offspring.

There are Ethical Issues to do with Human Cloning

1) The use of **human embryos** as a source of stem cells is **controversial**. The embryos are usually **destroyed** after the embryonic stem cells have been harvested — some people believe that doing this is **destroying a human life**.
2) Some people think a **cloned human** would have a **lower quality of life**, e.g. they might suffer **social exclusion** or have difficulty developing their own **personal identity**.
3) Some people think that cloning humans would be **wrong** because it **undermines** natural **sexual reproduction**, and traditional **family structures**.

Reproductive cloning of humans is currently illegal in the UK.

Plants can be Artificially Cloned using Tissue Culture

1) **Plants** can be **cloned** from existing plants using a technique called **tissue culture**. Here's how it's done:

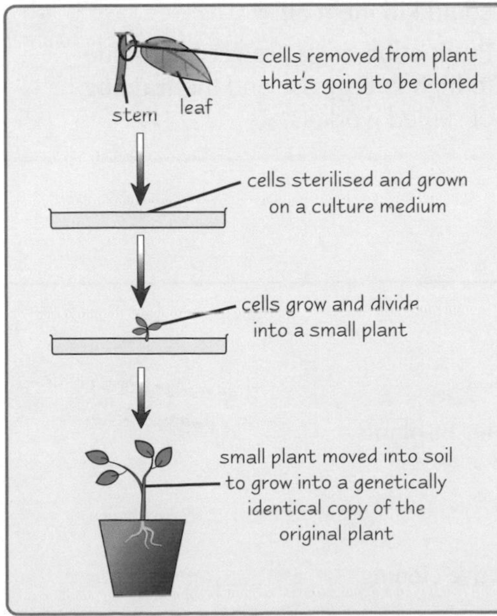

1) **Cells** are taken from the original plant that's going to be cloned.
2) Cells from the **stem** and **root tips** are used because they're **stem cells** — like in humans, plant stem cells can develop into **any type of cell**.
3) The cells are **sterilised** to kill any **microorganisms** — bacteria and fungi **compete** for nutrients with the **plant cells**, which **decreases** their **growth rate**.
4) The cells are placed on a **culture medium** containing plant **nutrients** (like **glucose** for **respiration**) and **growth factors** (such as **auxins**).
5) When the cells have **divided** and **grown** into a **small plant** they're taken out of the medium and **planted in soil** — they'll develop into plants that are **genetically identical** to the **original plant**.

2) Tissue culture is used to clone plants that **don't readily reproduce** or are **endangered** or **rare**, e.g. British orchids.
3) It's also used to grow **whole plants** from **genetically engineered plant cells**.
4) **Micropropagation** is when tissue culture is used to produce **lots** of cloned plants **very quickly**. **Cells** are taken from developing cloned plants and **subcultured** (grown on another fresh culture medium) — **repeating** this process creates **large numbers** of clones.

Cloning

Some Plants can Produce Natural Clones by Vegetative Propagation

Vegetative propagation is the natural production of plant clones from **non-reproductive tissues**, e.g. roots, leaves and stems. Plants grow **structures** on roots, leaves or stems that can **grow** into an identical **new plant**. You need to know how **elm trees** produce clones from structures called **suckers**:

1) A sucker is a **shoot** that grows from the **shallow roots** of an elm tree.
2) Suckers grow from **sucker buds** (undeveloped shoots) that are scattered around the tree's **root system**. The buds are **normally dormant**.
3) During times of **stress** (e.g. drought, damage or disease) or when a tree is **dying**, the **buds** are **activated** and suckers begin to form.
4) Suckers can pop up many metres **away** from the parent tree, which can help to **avoid** the **stress** that triggered their growth.
5) They eventually form completely **separate trees** — **clones** of the tree that the suckers grew from.

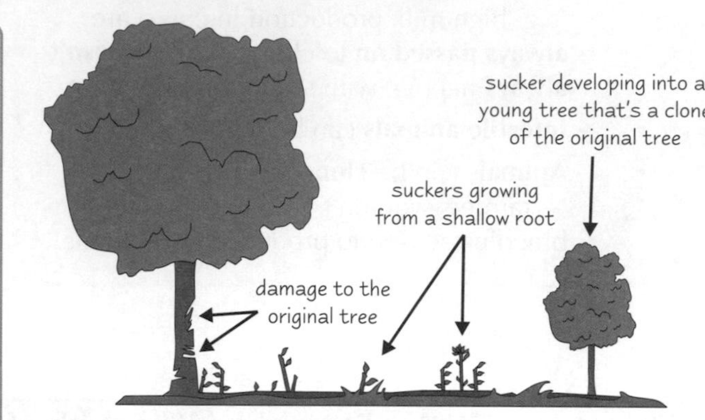

Plant Cloning in Agriculture has Advantages and Disadvantages

Advantages

- **Desirable genetic characteristics** (e.g. high fruit production) are **always passed on** to clones. This **doesn't always** happen when plants **reproduce sexually**.
- Plants can be reproduced in **any season** because tissue culture (see previous page) is carried out **indoors**.
- **Sterile plants** can be **reproduced**.
- Plants that take a **long time** to produce **seeds** can be **reproduced quickly**.

Disadvantages

- **Undesirable genetic characteristics** (e.g. producing fruit with lots of seeds) are **always passed on** to clones.
- **Cloned plant populations** have **no genetic variability**, so a **single disease** could **kill** them all.
- **Production costs** are **very high** due to **high energy use** and the **training** of skilled workers.

Practice Questions

Q1 What type of cells are made by non-reproductive cloning?

Q2 Name the technique that can be used to produce artificial clones of plants.

Q3 Give one disadvantage of plant cloning in agriculture.

These questions cover pages 70-72.

Exam Question

Q1 Scientists in the UK are using stem cells produced by non-reproductive cloning to research treatments for diseases like Parkinson's disease.

a) How does reproductive cloning differ from non-reproductive cloning? [2 marks]

b) Briefly describe the technique they might use to carry out non-reproductive cloning. [6 marks]

I ain't makin' no cloned elm tree, sucker...

Although it would be nice to have lots of clones doing your revision, exams and PE lessons, it's not going to happen. Sadly there's only one of you, and you need to learn about the different types of cloning, how they're done and their advantages and disadvantages. There are ethical issues with human cloning too — it's not everyone's cup of tea...

Biotechnology

The global biotechnology industry is humongous, but fortunately you've only got to learn three pages about it...

Biotechnology** is the **Use** of **Living Organisms in Industry

1) **Biotechnology** is the **industrial use** of **living organisms** to produce **food**, **drugs** and **other products**, e.g. yeast is used to make wine.
2) The living organisms used are mostly **microorganisms** (bacteria and fungi). Here are a few reasons why:

- Their **ideal growth conditions** can be **easily** created.
- They grow **rapidly** under the right conditions, so **products** can be made **quickly**.
- They can grow on a **range** of **inexpensive** materials.
- They can be grown at **any time** of the year.

3) Biotechnology also **uses parts** of **living organisms** (such as **enzymes**) to make products, e.g. rennet (a mix of enzymes) is extracted from calf stomachs and used to make cheese.
4) Enzymes used in industry can be **contained within the cells** of organisms — these are called **intracellular enzymes**.
5) Enzymes are also used that **aren't contained within cells** — these are called **isolated enzymes**. **Some are secreted naturally** by microorganisms (called **extracellular enzymes**), but others have to be **extracted**.
6) **Naturally secreted** enzymes are **cheaper** to use because it can be **expensive** to **extract** enzymes from cells.

Hooray, the rennet extractor's here.

Isolated Enzymes** can be **Immobilised

1) **Isolated enzymes** used in industry can become **mixed in** with the **products** of a reaction.
2) The **products** then need to be **separated** from this mixture, which can be **complicated** and **costly**.
3) This is **avoided** in large-scale production by using **immobilised enzymes** — enzymes that are **attached** to an **insoluble material** so they **can't** become mixed with the products.
4) There are **three main ways** that enzymes are **immobilised**:

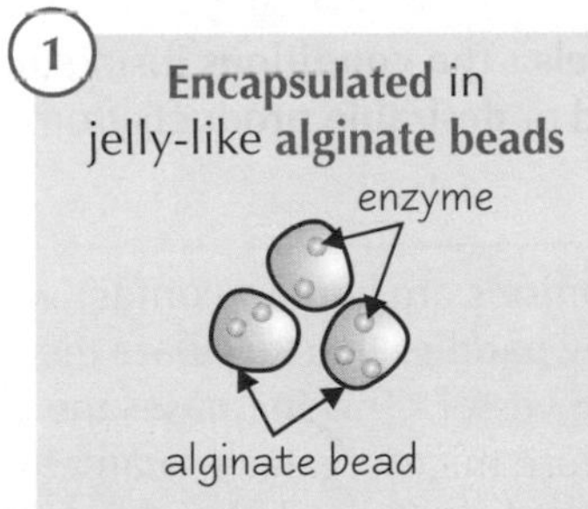

2 **Trapped** in a **silica gel matrix**

silica gel matrix

enzyme

3 **Covalently bonded** to **cellulose or collagen fibres**

cellulose or collagen fibres

enzyme

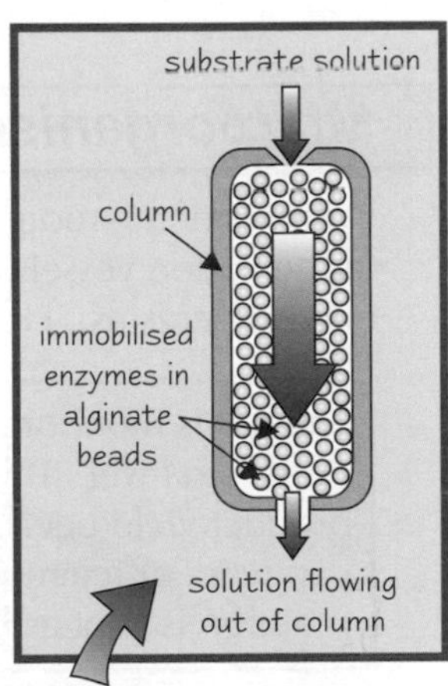

5) In industry, the **substrate solution** for a reaction is run through a **column** of **immobilised enzymes**.
6) The **active sites** of the enzymes are still **available** to **catalyse** the reaction but the solution flowing **out of** the column will **only** contain the **desired product**.
7) Here are some of the **advantages** of using **immobilised enzymes** in industry:

- Columns of immobilised enzymes can be **washed** and **reused** — this **reduces** the **cost** of running a reaction on an **industrial scale** because you don't have to **keep buying** new enzymes.
- The product **isn't mixed** with the enzymes — **no money** or **time** is **spent** separating them out.
- Immobilised enzymes are **more stable** than free enzymes — they're less likely to **denature** (become inactive) in **high temperatures** or **extremes** of **pH**.

Biotechnology

Closed Cultures of Microorganisms follow a Standard Growth Curve

1) A **culture** is a **population** of one type of microorganism that's been grown under **controlled conditions**.
2) A **closed culture** is when growth takes place in a vessel that's **isolated** from the **external environment** — extra nutrients **aren't added** and waste products **aren't removed** from the vessel **during growth**.
3) In a closed culture a population of microorganisms follows a **standard growth curve**:

(1) **Lag phase** — the population size **increases slowly** because the **microorganisms** have to make enzymes and other molecules before they can reproduce. This means the **reproduction rate** is **low**.

(2) **Exponential phase** — the population size **increases quickly** because the culture **conditions** are at their **most favourable** for **reproduction** (**lots of food** and **little competition**). The number of microorganisms **doubles** at **regular intervals**.

(3) **Stationary phase** — the population size **stays level** because the **death rate** of the microorganisms **equals** their **reproductive rate**. Microorganisms **die** because there's **not enough food** and poisonous **waste products build up**.

(4) **Decline phase** — the population size **falls** because the **death rate** is **greater** than the **reproductive rate**. This is because food is very **scarce** and waste products are at **toxic levels**.

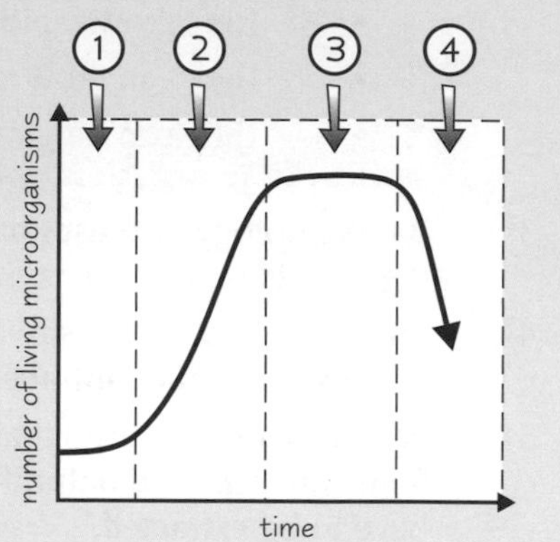

4) When growing conditions are **favourable** (e.g. during the **exponential phase**) microorganisms produce **primary metabolites** — **small molecules** that are **essential** for the **growth** of the microorganisms.
5) When growing conditions are **less favourable** (e.g. during the **stationary phase**) some microorganisms produce **secondary metabolites** — molecules that **aren't essential** for **growth** but are **useful** in **other ways**.
6) **Secondary metabolites** help microorganisms **survive**, e.g. the **antibiotic penicillin** is a secondary metabolite made by ***Penicillium*** (a fungus). It **kills bacteria** that **inhibit** its **growth**.
7) Some secondary metabolites are **desirable** to **biotechnology industries**, e.g. *Penicillium* is **cultured** on an **industrial scale** to produce lots of penicillin — it's used to treat **bacterial infections** in humans and animals.

Microorganisms are Grown in Fermentation Vessels

Cultures of microorganisms are grown in **large containers** called **fermentation vessels**. The **conditions** inside the fermentation vessels are kept at the **optimum for growth** — this **maximises** the **yield** of **desirable products** from the microorganisms. Here's a bit about how they work:

The **pH** is **monitored** and kept at the **optimum level**. This **increases** the product yield because **enzymes** can work **efficiently**, so the **rate of reaction** is kept as **high** as possible.

The **temperature** is kept at the **optimum level** by a **water jacket** that surrounds the vessel. This **increases** the product yield because **enzymes** can work **efficiently**, so the **rate of reaction** is kept as **high** as possible.

waste gas out e.g. CO_2
paddles
pH probe
water in
water out
water jacket
culture medium
sterile air in
product

Microorganisms are kept in **contact** with **fresh medium** by **paddles** that **circulate** the medium around the vessel. This **increases** the product yield because microorganisms can **always access** the **nutrients** needed for **growth**.

The volume of **oxygen** is kept at the **optimum level** for **respiration** by pumping in sterile air when needed. This **increases** the product yield because microorganisms can always **respire** to provide the **energy** for **growth**.

Vessels are **sterilised** between uses with **superheated steam** to kill any **unwanted organisms**. This **increases** the product yield because the microorganisms **aren't competing** with other organisms.

Biotechnology

There are **Two Main Culture Methods — Batch** *and* **Continuous**

1) **Batch culture** is where microorganisms are grown in **individual batches** in a fermentation vessel — when one culture **ends** it's **removed** and then a **different batch** of microorganisms is grown in the vessel.
2) **Continuous culture** is where microorganisms are **continually grown** in a fermentation vessel **without stopping**.
3) Here are some of the **differences** between batch culture and continuous culture:

Batch Culture	Continuous Culture
A fixed volume of growth medium (nutrients) is added to the fermentation vessel at the start of the culture and no more is added. The culture is a closed system.	Growth medium flows through the vessel at a steady rate so there's a constant supply of fresh nutrients. The culture is an open system.
Each culture goes through the lag, exponential and stationary growth phases.	The culture goes through the lag phase but is then kept at the exponential growth phase.
The product is harvested once, during the stationary phase.	The product is continuously taken out of the fermentation vessel at a steady rate.
The product yield is relatively low — stopping the reaction and sterilising the vessel between fermentations means there's a period when no product is being harvested.	The product yield is relatively high — microorganisms are constantly growing at an exponential rate.
If contamination occurs it only affects one batch. It's not very expensive to discard the contaminated batch and start a new one.	If the culture is contaminated the whole lot has to be discarded — this is very expensive when the cultures are done on an industrial scale.
Used when you want to produce secondary metabolites.	Usually used when you want primary metabolites or the microorganisms themselves as the desired product.

Asepsis *is* **Important** *when* **Culturing Microorganisms**

1) **Asepsis** is the practice of **preventing contamination** of cultures by **unwanted microorganisms**.
2) It's important when culturing microorganisms because contamination can **affect** the **growth** of the microorganism that you're **interested in**.
3) Contaminated cultures in **laboratory experiments** give **inaccurate results**.
4) Contamination on an **industrial scale** can be **very costly** because **entire cultures** may have to be **thrown away**.
5) A number of **aseptic techniques** can be used when working with microorganisms:

- **Work surfaces** are **regularly disinfected** to minimise contamination.
- **Gloves should be worn** and **long hair** is **tied back** to prevent it from falling into anything.
- The **instruments** used to **transfer** cultures are **sterilised before** and **after** each use, e.g. **inoculation loops** (small wire loops) are **heated** using a **Bunsen burner** to **kill** any microorganisms on them.
- In laboratories, the **necks** of **culture containers** are **briefly flamed** before they're **opened** or **closed** — this causes **air** to **move out** of the container, **preventing** unwanted microorganisms from **falling in**.
- **Lids** are **held over** open containers after they're removed, instead of putting them on a work surface. This **prevents** unwanted microorganisms from **falling** onto the culture.

Practice Questions

Q1 Give one way that enzymes are immobilised.

Q2 Why is it important to maintain the pH level in a fermentation vessel?

Q3 Why is asepsis important when culturing microorganisms?

These questions cover pages 73-75.

Exam Question

Q1 Describe and explain the standard growth curve of microorganisms in a closed culture. [8 marks]

Calf stomachs, yeast and sterile conditions — biotechnology is sexy stuff...

Wow, biology and technology fused together — forget bionic arms, legs and eyes though, growing bacteria in a tank is where it's at. Just think of yourself like an immobilised enzyme in a column — the substrate going in is the information on these pages, then all the desired information will flow out of you (not as a liquid hopefully) onto the exam paper.

Common Techniques

This section is all about technologies used to investigate and fiddle about with genes. So get your deerstalker hat on and your magnifying glass out...

Gene Technologies — Techniques Used to Study Genes

Gene technologies are basically all the **techniques** used to **study genes** and their **function** — you need to learn some of these techniques for the exam. They include:

- The **polymerase chain reaction** (**PCR**) (see below).
- Cutting out DNA fragments using **restriction enzymes** (see next page).
- **Gel electrophoresis** (see next page).
- Finding specific sequences of DNA using **DNA probes** (see p. 83).

As well as helping us to study genes, these techniques have **other uses**, such as in **genetic engineering** (see p. 78) and **gene therapy** (see p. 82).

'I know all about jean technology, baby...'

Multiple Copies of a DNA Fragment can be Made Using PCR

The **polymerase chain reaction** (PCR) can be used to make **millions of copies** of a fragment of DNA (containing the gene or bit of DNA you're interested in) in just a few hours. PCR has **several stages** and is **repeated** over and over to make lots of copies:

1) A reaction mixture is set up that contains the **DNA sample**, **free nucleotides**, **primers** and **DNA polymerase.**
   - **Primers** are short pieces of DNA that are **complementary** to the bases at the **start** of the fragment you want.
   - **DNA polymerase** is an **enzyme** that creates new DNA strands.
2) The DNA mixture is **heated** to **95 °C** to break the **hydrogen bonds** between the two strands of DNA.
3) The mixture is then **cooled** to between **50** and **65 °C** so that the primers can **bind** (**anneal**) to the strands.

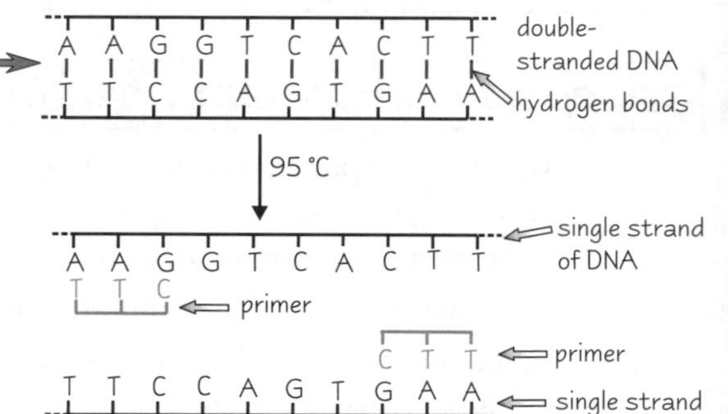

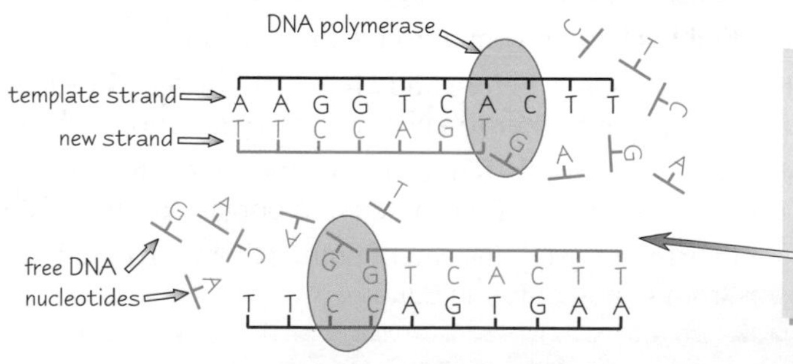

4) The reaction mixture is heated to **72 °C**, so **DNA polymerase** can **work**.
5) The DNA polymerase **lines up** free DNA nucleotides **alongside** each **template strand.** Complementary **base pairing** means **new complementary strands** are formed.

6) **Two new copies** of the fragment of DNA are formed and **one cycle** of PCR is **complete**.
7) The cycle starts again, with the mixture being heated to 95 °C and this time **all four strands** (two original and two new) are used as **templates**.
8) Each PCR cycle **doubles** the amount of DNA, e.g. **1st cycle** = 2 × 2 = **4 DNA fragments**, **2nd cycle** = 4 × 2 = **8 DNA fragments**, **3rd cycle** = 8 × 2 = **16 DNA fragments**, and so on.

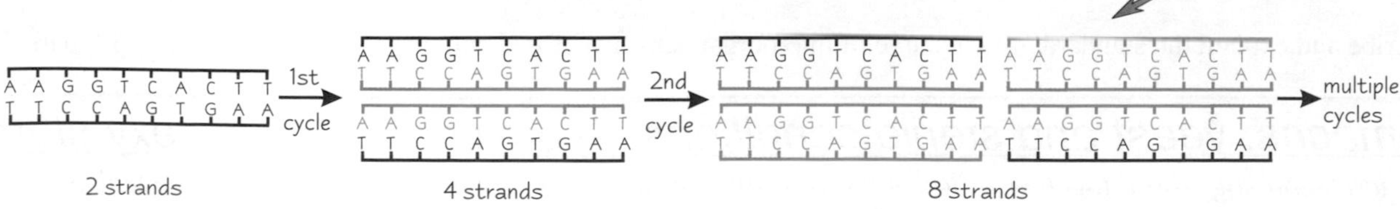

Common Techniques

Restriction Enzymes can be Used to Cut Out DNA Fragments

You can get a DNA fragment from an organism's DNA by using **restriction enzymes**:

1) Some sections of DNA have **palindromic** sequences of **nucleotides**. These sequences consist of **antiparallel base pairs** (base pairs that read the **same** in **opposite directions**). 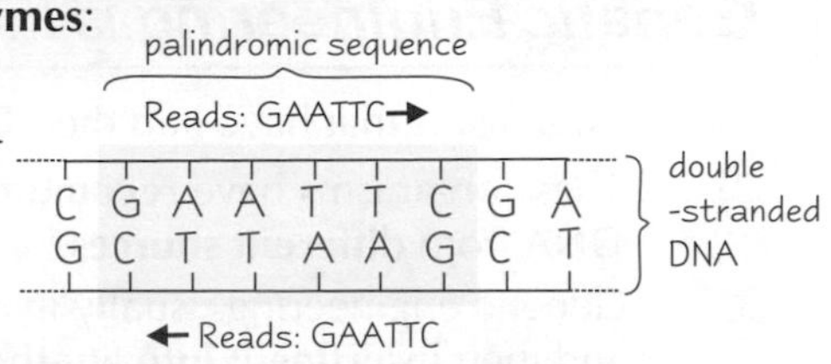
2) **Restriction enzymes** are enzymes that **recognise specific** palindromic sequences (known as **recognition sequences**) and **cut** (**digest**) the DNA at these places.
3) Different restriction enzymes cut at **different specific** recognition sequences, because the **shape** of the recognition sequence is **complementary** to an enzyme's **active site**. E.g. the restriction enzyme *Eco*RI cuts at GAATTC, but *Hind*III cuts at AAGCTT.
4) If recognition sequences are present at **either side** of the DNA fragment you want, you can use restriction enzymes to **separate** it from the rest of the DNA.
5) The DNA sample is **incubated** with the specific restriction enzyme, which **cuts** the DNA fragment out via a **hydrolysis reaction**.
6) Sometimes the cut leaves **sticky ends** — **small tails** of **unpaired bases** at **each end** of the fragment. Sticky ends can be used to **bind** (**anneal**) the DNA fragment to another piece of DNA that has sticky ends with **complementary sequences**.

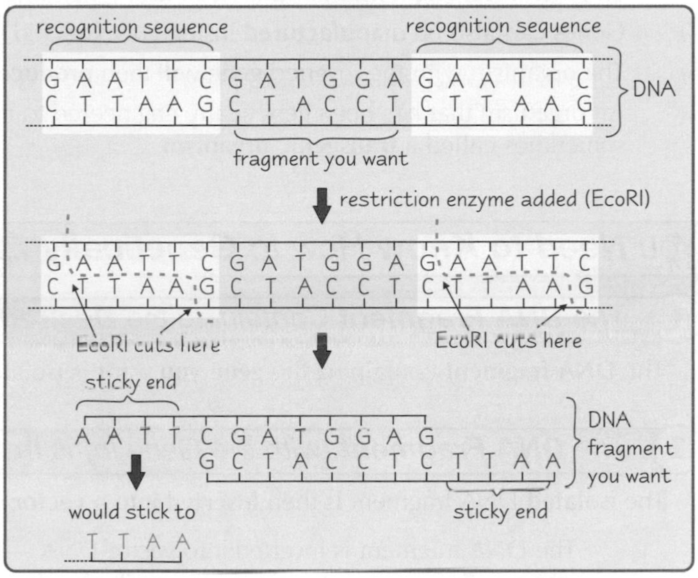

Electrophoresis Separates DNA Fragments by Size

1) A **fluorescent tag** is added to all the DNA fragments so they can be viewed under **UV light**.
2) The DNA is placed into a **well** in a slab of **gel** and covered in a **buffer solution** that **conducts electricity**.
3) An **electrical current** is passed through the gel — DNA fragments are **negatively charged**, so they **move towards** the **positive electrode** at the far end of the gel (called the **anode**).
4) **Small** DNA fragments move **faster** and **travel further** through the gel, so the DNA fragments **separate** according to **size**.
5) The DNA fragments are viewed as **bands** under **UV light**.

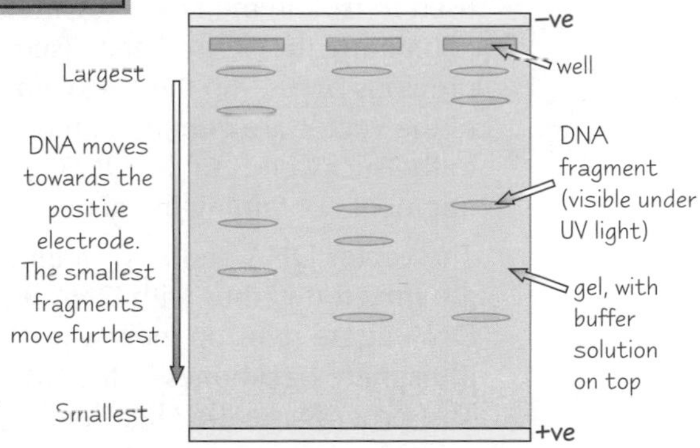

The **size** of a DNA fragment is **measured** in **bases**, e.g. ATCC = 4 bases or base pairs, **1000 bases** is **one kilobase** (1 kb).

Practice Questions

Q1 What are gene technologies?

Q2 What are restriction enzymes?

Exam Questions

Q1 Describe and explain how to produce multiple copies of a DNA fragment using PCR. [6 marks]

Q2 Describe and explain how electrophoresis works. [5 marks]

Sticky ends — for once a name that actually makes sense.

Gene technologies aren't the sort of technology you can buy in all good electrical stores, but they're still quite cool. Take PCR for example — you can throw in a DNA fragment and get out squillions of copies of it. Just like that. Amazing.

Genetic Engineering

Genetic engineering — you need to know what it is and how it's done... (unlucky)...

Genetic Engineering is the Manipulation of an Organism's DNA

1) Organisms that have had their **DNA altered** by genetic engineering are called **transformed organisms.**
2) These organisms have **recombinant DNA** — **DNA** formed by **joining together** DNA from **different sources.**
3) Genetic engineering usually involves **extracting** a **gene** from **one organism** and then **inserting** it **into another organism** (often one that's a **different species**).
4) Genes can also be **manufactured** instead of extracted from an organism.
5) The organism with the inserted gene will then **produce the protein** coded for by that gene.
6) An organism that has been genetically engineered to include a **gene** from a **different species** is sometimes called a **transgenic organism.**

Transformed organisms are also known as genetically engineered or genetically modified organisms.

You Need to Know How to Genetically Engineer a Microorganism

1) The DNA Fragment Containing the Desired Gene is Obtained

The **DNA fragment** containing the **gene you want** is isolated using **restriction enzymes** (see previous page).

2) The DNA Fragment (with the Gene in) is Inserted into a Vector

The **isolated** DNA fragment is then **inserted into** a **vector** using **restriction enzymes** and **ligase** (an enzyme):

1) The DNA fragment is inserted into vector DNA — a **vector** is something that's used to **transfer DNA** into a **cell**. They can be **plasmids** (**small, circular molecules** of DNA in **bacteria**) or **bacteriophages** (**viruses** that **infect** bacteria).
2) The vector DNA is **cut open** using the **same** restriction enzyme that was used to **isolate** the DNA fragment containing the desired gene (see previous page). So the **sticky ends** of the vector are **complementary** to the sticky ends of the DNA fragment containing the gene.
3) The vector DNA and DNA fragment are **mixed together** with **DNA ligase**. DNA ligase **joins** up the **sugar-phosphate backbones** of the two bits. This process is called **ligation**.
4) The new combination of bases in the DNA (vector DNA + DNA fragment) is called **recombinant DNA**.

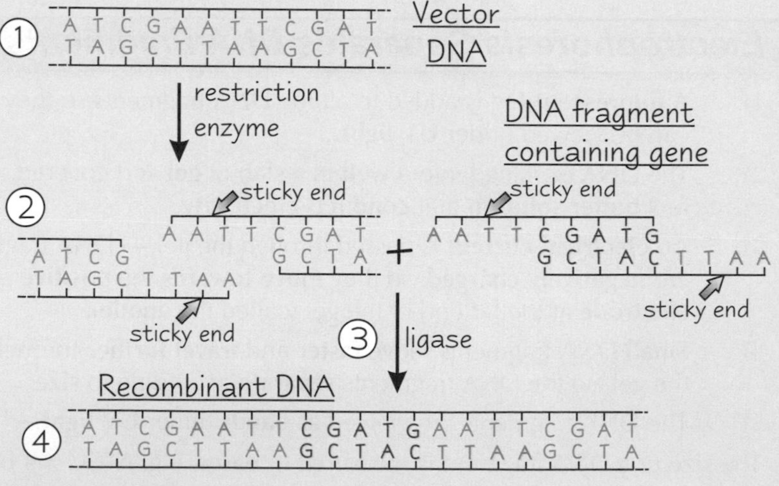

3) The Vector Transfers the Gene into the Bacteria

1) The **vector** with the **recombinant DNA** is used to **transfer** the gene into the **bacterial cells.**
2) If a **plasmid vector** is used, the bacterial cells have to be **persuaded** to **take in** the plasmid vector and its DNA. E.g. they're placed into ice-cold **calcium chloride** solution to make their cell walls more **permeable**. The **plasmids** are **added** and the mixture is **heat-shocked** (heated to around **42 °C** for **1-2 minutes**), which encourages the cells to take in the plasmids.
3) With a **bacteriophage** vector, the bacteriophage will **infect** the bacterium by **injecting** its **DNA** into it. The phage DNA (with the desired gene in it) then **integrates** into the bacterial DNA.
4) **Cells** that **take up** the vectors containing the desired gene are genetically engineered, so are called **transformed**.

Genetic Engineering

4 Identify the Transformed Bacteria

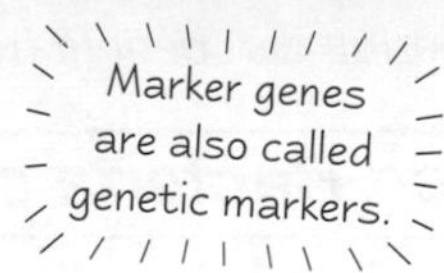

Not all the bacteria will have **taken up** the vector.
Marker genes can be used to **identify** the ones that **have**:

1) **Marker genes** can be inserted into vectors at the **same time** as the desired gene. This means any **transformed bacterial cells** will contain the desired gene **and** the marker gene.
2) The bacteria are **grown** on **agar plates** and each cell **divides** and **replicates** its DNA, creating a **colony** of **cells**.
3) Transformed cells will produce colonies where **all the cells** contain the desired gene and the marker gene.
4) The marker gene can code for **antibiotic resistance** — bacteria are grown on agar plates **containing** the **antibiotic**, so **only** cells that have the **marker gene** will **survive** and **grow**.
5) The marker gene can code for **fluorescence** — when the agar plate is placed under a **UV light only** transformed cells will **fluoresce**.

It's Useful for Microorganisms to be Able to Take Up Plasmids

Microorganisms can **take up plasmids** from their surroundings, which is **beneficial** because the plasmids often contain **useful genes**. This means the microorganisms gain **useful characteristics**, so they're more likely to have an **advantage** over other microorganisms, which **increases** their **chance** of **survival**. Plasmids may contain:

- Genes that code for **resistance** to **antibiotics**, e.g. genes for enzymes that **break down antibiotics**.
- Genes that help microorganisms **invade hosts**, e.g. genes for enzymes that **break down host tissues**.
- Genes that mean microorganisms can use **different nutrients**, e.g. genes for enzymes that break down **sugars** not normally used.

Practice Questions

Q1 What is the name for an organism that has had its DNA altered?

Q2 What is a vector?

Q3 Other than a plasmid, give an example of a vector.

Q4 Name the type of enzyme that can be used to cut DNA.

Q5 What is the name of the type of DNA formed from vector DNA and an inserted DNA fragment?

Q6 What is a marker gene?

Agar plate — agar plate, colony A, colony B
Agar plate with penicillin

Exam Question

Q1 A scientist has genetically engineered some bacterial cells to contain a desired gene and a gene that gives resistance to penicillin. The cells were grown on an agar plate and then transferred to a plate containing penicillin. The two plates are shown above.

a) Explain why the scientist thinks colony A contains transformed bacterial cells, but colony B doesn't. [2 marks]

b) Explain how the scientist might have inserted the desired gene into the plasmid. [3 marks]

c) Explain why being able to take up plasmids is useful to bacteria. [2 marks]

Examiners — genetically engineered to contain marker genes...

This stuff might seem tricky the first time you read it, but it's not too bad really — you get the gene you want and bung it in a vector, the vector gets the gene into the cell (it's kind of like a delivery boy), then all you have to do is figure out which cells have got the gene. Easy peasy. Unfortunately you need to know each stage in detail, so get learnin'.

Genetic Engineering

Genetic engineering can benefit humans in loads of different ways...

Transformed **Bacteria** can be used to **Produce Human Insulin**

See page 18 for more on Type 1 diabetes.

People with **Type 1 diabetes** need to **inject insulin** to **regulate** their **blood glucose concentration**. Insulin used to be obtained from the **pancreases** of dead animals, such as **pigs**. Nowadays we use **genetically engineered bacteria** to manufacture **human insulin**. Here's how the whole process works:

1) The **gene** for **human insulin** is **identified** and **isolated** using **restriction enzymes**.
2) A **plasmid** is **cut open** using the **same** restriction enzyme that was used to isolate the insulin gene.
3) The **insulin gene** is **inserted** into the **plasmid** (forming **recombinant DNA**).
4) The plasmid is **taken up** by bacteria and any **transformed** bacteria are **identified** using **marker genes**.
5) The bacteria are **grown** in a **fermenter** — **human insulin** is **produced** as the bacteria **grow** and **divide**.
6) The human insulin is **extracted** and **purified** so it can be **used in humans**.

These techniques are covered in more detail on page 78.

There are many **advantages** of using **genetically engineered human insulin** over **animal insulin**:

- It's **identical** to the insulin in our bodies, so it's **more effective** than animal insulin and there's **less risk** of an **allergic reaction**.
- It's **cheaper** and **faster** to produce than animal insulin, providing a **more reliable** and **larger supply** of insulin.
- Using genetically engineered insulin **overcomes** any **ethical** or **religious issues** arising from using animal insulin.

Transformed **Plants** can be Used to **Reduce Vitamin Deficiency**

Golden Rice is a type of **genetically engineered rice**. The rice is genetically engineered to contain a **gene** from a **maize plant** and a **gene** from a **soil bacterium**, which together enable the rice to produce **beta-carotene**. The beta-carotene is used by our bodies to produce **vitamin A**. *Golden Rice* is being developed to **reduce vitamin A deficiency** in areas where there's a **shortage** of **dietary vitamin A**, e.g. south Asia, parts of Africa. Here's how *Golden Rice* is produced:

1) The ***psy*** gene (from maize) and the ***crtl*** gene (from the soil bacterium) are **isolated** using **restriction enzymes**.
2) A **plasmid** is **removed** from the ***Agrobacterium tumefaciens*** **bacterium** and **cut open** with the **same** restriction enzymes.
3) The *psy* and *crtl* genes and a **marker gene** are **inserted** into the plasmid.
4) The **recombinant plasmid** is **put back into** the bacterium.
5) **Rice plant cells** are incubated with the **transformed** *A. tumefaciens* bacteria, which **infect** the rice plant cells.
6) *A. tumefaciens* **inserts** the **genes** into the **plant cells' DNA**, creating **transformed rice plant cells**.
7) The rice plant cells are then grown on a **selective medium** — only transformed rice plants will be able to **grow** because they contain the marker gene that's needed to grow on this medium.

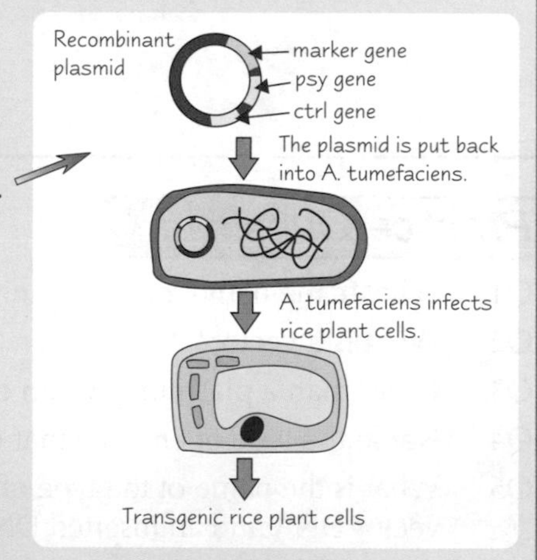

Transformed **Animals** can be Used to **Produce Organs** for **Transplant**

1) **Organ failure** (e.g. kidney or liver failure) may be **treated** with an **organ transplant**.
2) However, there's a **shortage** of **donor organs** available for transplant in the UK, which means many people **die** whilst **waiting** for a suitable donor organ.
3) **Xenotransplantation** is the **transfer** of **cells**, **tissues** or **organs** from **one species** to **another**.
4) It's hoped that xenotransplantation can be used to provide **animal donor organs** for **humans**.
5) With any form of transplantation there's a chance of **rejection** — the **immune system** of the **recipient recognises proteins** on the **surface** of the transplanted cells as **foreign** and starts an **immune response** against them.
6) Rejection is an **even greater** problem with xenotransplantation because the **genetic differences** between organisms of **different species** are even **greater** than between organisms of the same species.

Genetic Engineering

7) Scientists are trying to **genetically engineer animals** so that their **organs aren't rejected** when transplanted into humans. Here's how:

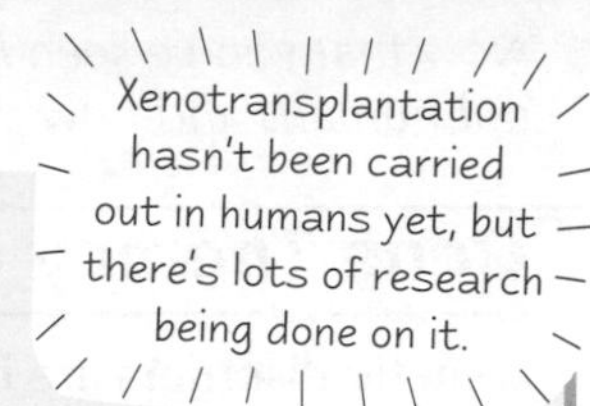

1) Genes for HUMAN cell-surface proteins are INSERTED into the animal's DNA:

Human genes for **human cell-surface proteins** are **injected** into a **newly fertilised animal embryo**. The genes **integrate** into the **animal's DNA**. The animal then **produces human cell-surface proteins**, which reduces the risk of transplant rejection.

2) Genes for ANIMAL cell-surface proteins are 'KNOCKED OUT' — removed or inactivated:

- Animal genes involved in making cell-surface proteins are **removed** or **inactivated** in the **nucleus** of an **animal cell**. The nucleus is then **transferred** into an **unfertilised animal egg cell** (this is called **nuclear transfer**). The egg cell is then **stimulated** to **divide** into an embryo and the animal created **doesn't produce animal** cell-surface proteins, which reduces the risk of transplant rejection.
- For example, **pigs** have a sugar called **Gal-alpha(1,3)-Gal** attached to their cell-surface proteins, which humans don't. Scientists have developed a **knockout pig** that **doesn't produce** the **enzyme** needed to **make** this sugar.

There are Some **Ethical Issues** Surrounding **Genetic Engineering**

Genetic engineering can be used for **loads of things** other than producing insulin, reducing vitamin A deficiency and producing organs suitable for transplant from animals. For example it can be used to produce **pest-** or **herbicide-resistant crops** and **drugs** (and could even be used to genetically engineer **humans**). All these applications have ethical issues and concerns surrounding them:

1) Some people are worried that using **antibiotic-resistance** genes as **marker genes** may **increase** the number of **antibiotic-resistant**, **pathogenic** (disease-causing) **microorganisms** in our environment.
2) **Environmentalists** are worried that GM crops (like *Golden Rice*) may encourage **farmers** to carry out **monoculture** (where only one type of crop is planted). Monoculture **decreases biodiversity** and could leave the **whole crop vulnerable** to **disease**, because all the plants are **genetically identical**.
3) Some people are worried that genetically engineering **animals** for **xenotransplantation** may **cause them suffering**.
4) Some people are concerned about the possibility of **'superweeds'** — weeds that are **resistant** to **herbicides** because they've bred with **genetically engineered herbicide-resistant crops**.
5) Some people are concerned that large biotechnology companies may use GM crops to **exploit farmers** in **poor countries** — e.g. by selling them crops that they **can't** really **afford**.
6) Some people worry **humans** will be genetically engineered (e.g. to be more intelligent), creating a **genetic underclass**. This is currently **illegal** though.

Practice Questions

Q1 Give two advantages of using human insulin produced by genetic engineering compared to using animal insulin.

Q2 What is xenotransplantation?

Q3 How could xenotransplantation benefit humans?

Q4 Give two ethical issues surrounding genetic engineering.

Exam Questions

Q1 People with Type 1 diabetes need to inject insulin to regulate their blood glucose concentration.
Describe how human insulin can be made using genetically engineered bacteria. [6 marks]

Q2 *Golden Rice* is a type of transformed rice. Outline the process used to create *Golden Rice*. [7 marks]

If only they could knockout the gene for smelly feet...

...or the gene for freckles... or spots... or a big nose... or chubby ankles... the list is endless. Not that I'm vain or anything. Anyway, make sure you know all the processes in detail — it's no good just knowing that Golden Rice *is a genetically engineered crop, you need to know how it was genetically engineered too. So knuckle down and go over the page again...*

Gene Therapy and DNA Probes

Now that you've seen how microorganisms, plants and animals can be genetically engineered it seems a shame to leave out humans and how gene therapy could be used to cure some disorders. Then you need to know all about DNA probes.

Gene Therapy Could be Used to Cure Genetic Disorders

Genetic disorders are **inherited disorders** caused by **abnormal genes** or **chromosomes**, e.g. cystic fibrosis. **Gene therapy** could be used to **cure** these disorders — it **isn't** being used yet but some treatments are undergoing **clinical trials**.

How it works:

1) Gene therapy involves **altering alleles** inside cells to cure **genetic disorders**.
2) How you do this depends on whether the genetic disorder is caused by a **dominant allele** or two **recessive alleles**:
   - If it's caused by two **recessive** alleles you can **add** a working **dominant allele** to make up for them.
   - If it's caused by a **dominant** allele you can **'silence'** the **dominant allele** (e.g. by sticking a bit of DNA in the middle of the allele so it doesn't work any more).

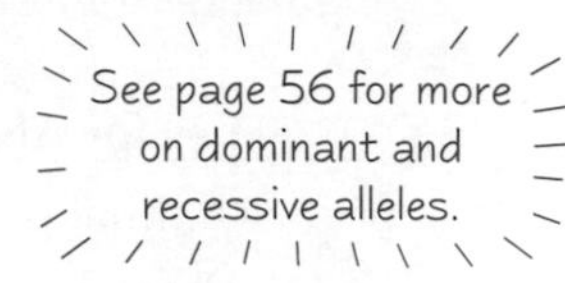
See page 56 for more on dominant and recessive alleles.

A DNA-filled doughnut — surely the best way to deliver new alleles...

How you get the 'new' allele (DNA) inside the cell:

1) The allele is **inserted into cells** using **vectors**.
2) Different **vectors** can be used, e.g. altered **viruses**, **plasmids** or **liposomes** (spheres made of lipid).

There are two types of gene therapy:

1) **Somatic therapy** — this involves **altering** the **alleles** in **body cells**, particularly the cells that are **most affected** by the disorder. For example, **cystic fibrosis** (CF) is a genetic disorder that's very **damaging** to the **respiratory system**, so somatic therapy for CF **targets** the epithelial cells lining the lungs. Somatic therapy doesn't affect the individual's **sex cells** (sperm or eggs) though, so any **offspring** could still **inherit** the disease.
2) **Germ line therapy** — this involves **altering** the **alleles** in the **sex cells**. This means that **every cell** of **any offspring** produced from these cells will be **affected** by the gene therapy and they **won't suffer from the disease**. Germ line therapy in humans is currently **illegal** though.

There are Advantages and Disadvantages to Gene Therapy

ADVANTAGES	DISADVANTAGES
It could prolong the lives of people with genetic disorders.	The effects of the treatment may be short-lived (only in somatic therapy).
It could give people with genetic disorders a better quality of life.	The patient might have to undergo multiple treatments (only in somatic therapy).
Carriers of genetic disorders might be able to conceive a baby without that disorder or risk of cancer (only in germ line therapy).	It might be difficult to get the allele into specific body cells.
It could decrease the number of people that suffer from genetic disorders (only in germ line therapy).	The body could identify vectors as foreign bodies and start an immune response against them.
	An allele could be inserted into the wrong place in the DNA, possibly causing more problems, e.g. cancer.
	An inserted allele could get overexpressed, producing too much of the missing protein.
	Disorders caused by multiple genes (e.g. cancer) would be difficult to treat with this technique.

There are also many **ethical issues** associated with gene therapy. For example, some people are worried that the technology could be used in ways **other** than for **medical treatment**, such as for treating the **cosmetic effects** of **aging**. Other people worry that there's the potential to do **more harm** than good by using the technology (e.g. risk of overexpression of genes — see table). There's also the concern that gene therapy is **expensive** — some people believe that **health service resources** could be **better spent** on other treatments that have passed clinical trials.

Gene Therapy and DNA Probes

DNA Probes can be used to Identify Specific Base Sequences in DNA

1) **DNA probes** (also called **gene probes**) can be used to **identify DNA fragments** that contain **specific sequences** of bases, e.g. they can be used to **locate genes** on chromosomes or see if a person's DNA **contains** a **mutated gene** (e.g. a gene that causes a genetic disorder).
2) DNA probes are **short strands** of **DNA**. They have a **specific base sequence** that's **complementary** to the target sequence — the specific sequence you're looking for.

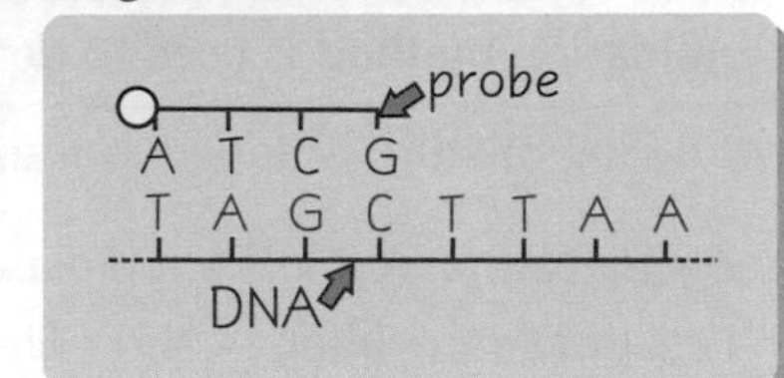

3) This means a DNA probe will **bind** (**hybridise**) to the **target sequence** if it's **present** in a **sample** of DNA.
4) A DNA probe also has a **label attached**, so that it can be **detected**. The two most common types of label are a **radioactive** label (detected using **X-ray film**) or a **fluorescent** label (detected using **UV light**).
5) For example, you can use a DNA probe to see if any members of a family have a **mutation** in a gene that causes a **genetic disorder**:
   - A **sample** of **DNA** from each family member is **digested** into fragments using **restriction enzymes** (see page 77) and **separated** using **electrophoresis** (see page 77).
   - The separated DNA fragments are then transferred to a **nylon membrane** and **incubated** with the **fluorescently labelled DNA probe**. The probe is **complementary** to the specific sequence of the mutated gene.
   - If the specific sequence **is present** in one of the DNA fragments, the DNA probe will **hybridise** (**bind**) to it.
   - The **membrane** is then **exposed** to **UV light** and if the specific sequence is present in one of the DNA fragments, then that band will **fluoresce** (**glow**).
   - For example, **person three** has a **visible band**, so that family member has the specific sequence **in** one of their DNA fragments, which means they **have** the **mutated gene**.

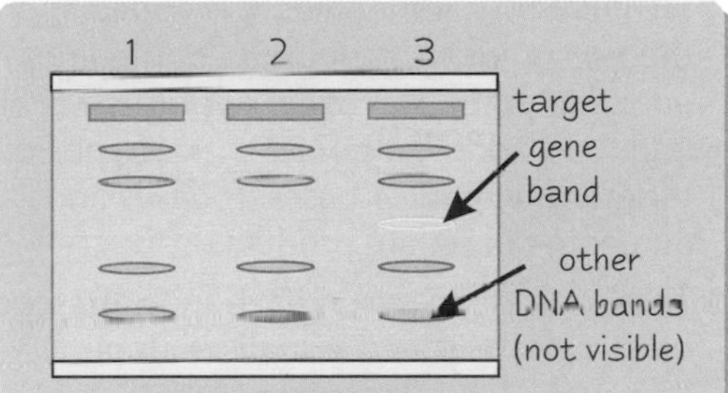

Practice Questions

Q1 How could gene therapy be used to supplement mutated recessive alleles?

Q2 How are supplementary alleles added to human DNA?

Q3 What does germ line gene therapy involve?

Q4 What is a DNA probe?

Q5 Name two types of label that can be attached to a DNA probe.

Exam Questions

Q1 A patient suffering from cystic fibrosis was offered gene therapy targeted at his lung epithelial cells to help treat the disease.

a) What does gene therapy involve? [1 mark]

b) What type of gene therapy was the patient offered? [1 mark]

Q2 Give three possible disadvantages of somatic gene therapy. [3 marks]

Q3 Erin has a family history of breast cancer. She has agreed to be screened for the mutated BRCA1 gene, which can cause breast cancer. A sample of her DNA is digested and separated using gel electrophoresis. Describe how a DNA probe could be used to identify the mutated BRCA1 gene. [4 marks]

DNA probes — don't worry, the DNA doesn't feel a thing...

You need to know exactly what gene therapy is and the differences between the two types ('somatic' means 'body', which should help you remember). How DNA probes work also needs to be etched in your memory — they can be used to tell if someone has a genetic disorder, but don't forget they can be used for other things too, like locating genes on chromosomes.

Sequencing Genes and Genomes

OK, these are the last two pages of the section, but sequencing is tricky business — so look sharp.

DNA can be Sequenced by the Chain-Termination Method

The **chain-termination method** is used to determine the **order** of **bases** in a section of **DNA** (gene):

1) The following mixture is added to **four separate** tubes:
   - A **single-stranded DNA template** — the DNA to sequence.
   - Lots of **DNA primer** — short pieces of DNA (see p. 76).
   - **DNA polymerase** — the enzyme that joins DNA nucleotides together.
   - **Free nucleotides** — lots of free A, T, C and G nucleotides.
   - **Fluorescently-labelled modified nucleotide** — like a normal nucleotide, but once it's added to a DNA strand, **no more** bases can be added after it. A **different** modified nucleotide is added to **each** tube (**A*, T*, C*, G***).
2) The tubes undergo **PCR**, which produces many **strands of DNA**. The strands are **different lengths** because each one **terminates** at a **different point** depending on where the modified nucleotide was added.
3) For example, in tube A (with the **modified adenine** nucleotide **A***) sometimes A* is **added** to the DNA at point 4 **instead** of A, **stopping** the **addition** of any more bases (the strand is **terminated**). Sometimes A is added at point 4, then **A*** is added at **point 5**. Sometimes A is added at **point 4**, A again at point 5, G at point 6 and **A*** is added at **point 7**. So strands of **three different lengths** (4 bases, 5 bases and 7 bases) all ending in **A*** are produced.
4) The DNA fragments in each tube are separated by **electrophoresis** and **visualised** under **UV light** (because of the **fluorescent label**).
5) The **complementary base sequence** can be **read** from the gel. The **smallest** nucleotide (e.g. one base) is at the **bottom** of the gel. Each band after this represents **one more base** added. So by reading the bands **from the bottom** of the gel **to the top**, you can build up the **DNA sequence** one base at a time.

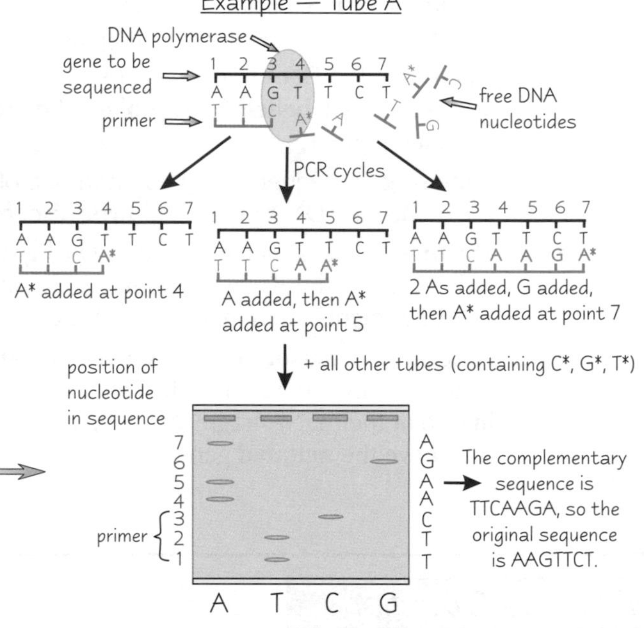

A Whole Genome can be Sequenced Using BACs

The **chain-termination method** can only be used for DNA fragments up to about **750 bp** long. So if you want to sequence the **entire genome** (all the DNA) of an organism, you need to chop it up into **smaller pieces** first. The smaller pieces are **sequenced** and then **put back in order** to give the sequence of the whole genome. Here's how it's done:

1) A genome is **cut** into **smaller fragments** (about 100 000 bp) using **restriction enzymes**.
2) The fragments are inserted into **bacterial artificial chromosomes** (**BACs**) — these are **man-made plasmids**. **Each** fragment in inserted into a **different BAC**.
3) The BACs are then **inserted** into **bacteria** — **each bacterium** contains a **BAC** with a **different DNA fragment**.
4) The bacteria **divide**, creating **colonies** of **cloned** (**identical**) cells that all contain a **specific DNA fragment**. Together the different colonies make a complete **genomic DNA library**.
5) **DNA** is **extracted** from **each colony** and **cut** up using restriction enzymes, producing **overlapping** pieces of DNA.
6) Each piece of DNA is **sequenced**, using the **chain-termination method**, and the pieces are **put back in order** to give the full sequence **from that BAC** (using **powerful computer systems**).
7) Finally the DNA fragments from **all the BACs** are **put back in order**, by computers, to **complete** the **entire genome**.

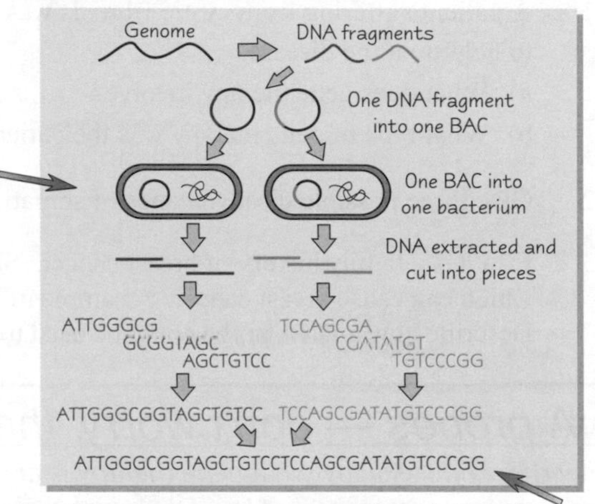

Sequencing Genes and Genomes

Sequenced Genomes can be Compared

Gene sequences and **whole genome** sequences can be compared **between** organisms of **different species** and between organisms of the **same species**. There are many reasons why we'd want to do this:

Look, when we stick our tongues out like this you can just TELL we're related, we don't need a genome comparison.

Comparing the genomes of DIFFERENT species can help us to:

1) Understand the **evolutionary relationships** between different species. **All** organisms **evolved** from **shared common ancestors** (relatives). **Closely related** species **evolved away** from each other more **recently** and so **share more DNA**. So DNA can tell us **how closely related** different species are. E.g. the genomes of **humans** and **chimpanzees** are about **94%** similar.
2) Understand the way in which **genes interact** during **development** and how they're **controlled**. For example, genome sequencing has shown that the **homeobox sequence** (see page 51) is the **same** in animals, plants and fungi. By studying how genes with the homeobox sequence work in the **Drosophila fruit fly** scientists can begin to piece together how they work in **humans** too.
3) Carry out **medical research. Human genes** that are associated with **disease**, like cancer or heart disease, can be found in the genomes of **other mammals**, such as mice and rats. This means mice or rats could be used as **animal models** for **research** into these diseases.

Comparing genomes of the SAME species can help us to:

1) Trace **early human migration**. For example, when different groups of early **humans separated** and **moved** to different parts of the world, their genomes **changed** in **slightly different ways**. By **comparing** the genomes of people from different parts of the world, it's possible to build up a picture of early human migration.
2) Study the **genetics** of **human diseases**. Some **gene mutations** have been **linked** to a **greater risk** of **disease** (e.g. mutations in the **BRCA1** gene are linked to **breast cancer**). Comparisons between the genomes of **sufferers** and **non-sufferers** can be used to **detect** particular **mutations** that could be responsible for the increased risk of disease.
3) Develop **medical treatments** for **particular genotypes**. The **same medicine** can be **more effective** in some patients than in others, which can be due to their **different genomes**. In the future, it may be possible to **sequence** a patient's genome so they can receive the **most effective medicine** for them.

Practice Questions

Q1 Give the name of a method used to sequence DNA.

Q2 What is a bacterial artificial chromosome?

Q3 Give two reasons why a scientist might want to compare the genomes of organisms from different species.

Q4 Give two reasons why a scientist might want to compare the genomes of organisms from the same species.

Exam Questions

Q1 To sequence a small DNA fragment, a single-stranded DNA template and DNA polymerase are needed.

a) Name the other three reactants needed for a sequencing reaction. [3 marks]

b) Describe and explain the process of sequencing a small DNA fragment. [6 marks]

Q2 The genomes of over 200 different species have been sequenced.
Describe how a genome can be sequenced using BACs. [8 marks]

Sequincing — so 80s...

Don't worry, the buzzing in your head is normal — it's due to information overload. So go get yourself a cuppa and a biccie and have a break. Then go over some of the difficult bits in this section again. Yes, I did say 'again'. But believe me, the more times you go over it the more things will click into place and you'll be even more prepared for the exam.

Ecosystems and the Nitrogen Cycle

All this ecology-type stuff is pretty wordy, so here are a nice few definitions to get you started. This way, you'll know what I'm banging on about throughout the rest of the section, and that always helps I think.

You Need to Learn Some Definitions to get you Started

Ecosystem	—	**All** the **organisms** living in a **particular area** and all the **non-living** (abiotic) conditions, e.g. a freshwater ecosystem such as a lake. Ecosystems are **dynamic systems** — they're **changing** all the time.
Habitat	—	The **place** where an organism **lives**, e.g. a rocky shore or a field.
Population	—	**All** the organisms of **one species** in a **habitat**.
Abiotic factors	—	The **non-living** features of the ecosystem, e.g. **temperature** and **availability of water**.
Biotic factors	—	The **living** features of the ecosystem, e.g. the presence of **predators** or **food**.
Producer	—	An organism that **produces organic molecules** using sunlight energy, e.g. plants.
Consumer	—	An organism that **eats other organisms**, e.g. animals and birds.
Decomposer	—	An organism that **breaks down dead** or **undigested organic material**, e.g. bacteria and fungi.
Trophic level	—	A **stage** in a **food chain** occupied by a particular **group** of organisms, e.g. producers are the first trophic level in a food chain.

Being a member of the undead made it hard for Mumra to know whether he was a living or a non-living feature of the ecosystem.

Energy is *Transferred Through Ecosystems*

1) The **main route** by which energy **enters** an ecosystem is **photosynthesis** (e.g. by plants, see p. 28). (Some energy enters sea ecosystems when bacteria use chemicals from deep sea vents as an energy source.)
2) During photosynthesis plants **convert sunlight energy** into a form that can be **used** by other organisms — plants are called **producers** (see table above).
3) Energy is **transferred** through the **living organisms** of an ecosystem when organisms **eat** other organisms, e.g. producers are eaten by organisms called **primary consumers**. Primary consumers are then eaten by **secondary consumers** and secondary consumers are eaten by **tertiary consumers**.
4) **Food chains** and **food webs** show how energy is **transferred** through an ecosystem.
5) **Food chains** show **simple lines** of energy transfer.
6) **Food webs** show **lots** of **food chains** in an ecosystem and how they **overlap**.
7) Energy locked up in the things that **can't be eaten** (e.g. bones, faeces) gets recycled back into the ecosystem by **decomposers**.

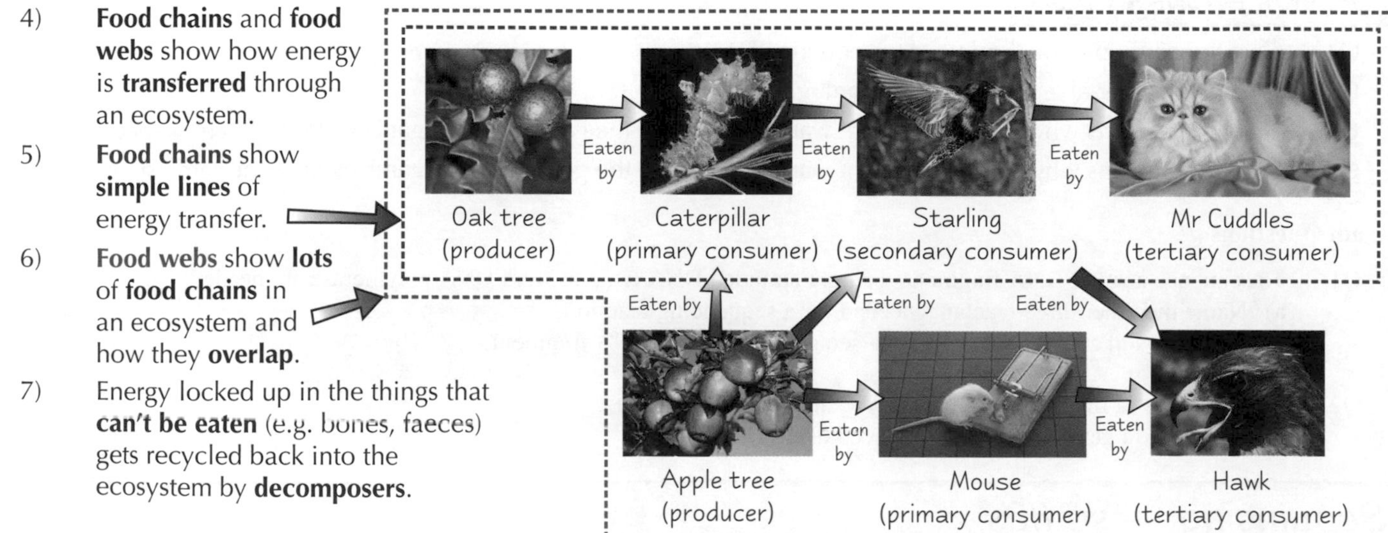

Ecosystems and the Nitrogen Cycle

The **Nitrogen Cycle** shows how **Nitrogen** is **Recycled** in **Ecosystems**

Plants and animals **need** nitrogen to make **proteins** and **nucleic acids** (DNA and RNA). The atmosphere's made up of about 78% nitrogen, but plants and animals **can't use it** in that form — they need **bacteria** to **convert** it into **nitrogen compounds** first. The **nitrogen cycle** shows how nitrogen is **converted** into a useable form and then **passed** on between different **living** organisms and the **non-living** environment.

The nitrogen cycle includes **food chains** (nitrogen is passed on when organisms are eaten), and four different processes that involve bacteria — **nitrogen fixation**, **ammonification**, **nitrification** and **denitrification**:

1 Nitrogen fixation

- **Nitrogen fixation** is when nitrogen **gas** in the atmosphere is turned into **ammonia** by **bacteria** called ***Rhizobium***. The ammonia can then be **used** by plants.
- *Rhizobium* are found inside **root nodules** (growths on the roots) of **leguminous** plants (e.g. peas, beans and clover).
- They form a **mutualistic** relationship with the plants — they provide the plant with **nitrogen compounds** and the plant provides them with **carbohydrates**.

2 Ammonification

- **Ammonification** is when nitrogen compounds from **dead organisms** are turned into **ammonium compounds** by **decomposers**.
- Animal **waste** (**urine** and **faeces**) also contains nitrogen compounds. These are also turned into ammonium compounds by decomposers.

3 Nitrification

- **Nitrification** is when **ammonium compounds** in the soil are **changed** into **nitrogen compounds** that can then be **used** by plants.
- First **nitrifying bacteria** called ***Nitrosomonas*** change **ammonium compounds** into **nitrites**.
- Then other nitrifying bacteria called ***Nitrobacter*** change **nitrites** into **nitrates**.

4 Denitrification

- **Denitrification** is when nitrates in the soil are **converted** into **nitrogen gas** by **denitrifying bacteria** — they use nitrates in the soil to carry out **respiration** and produce nitrogen gas.
- This happens under **anaerobic conditions** (where there's **no** oxygen), e.g. in **waterlogged** soils.

Other ways that **nitrogen** gets into an **ecosystem** is by **lightning** (which **fixes atmospheric nitrogen**) or by **artificial fertilisers** (they're **produced from atmospheric nitrogen** on an **industrial scale** in the **Haber process**).

Practice Questions

Q1 Define ecosystem.

Q2 What is a consumer?

Q3 Name the organisms that break down dead organic material.

Q4 What is nitrification?

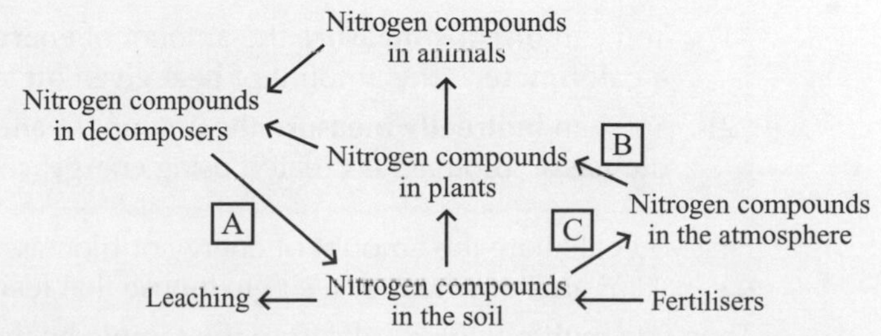

Exam Question

Q1 The diagram on the right shows the nitrogen cycle.

a) Name the processes labelled A and C in the diagram. [2 marks]

b) Name and describe process B in detail. [3 marks]

Nitrogen fixation — cheaper than a shoe fixation...

The nitrogen cycle's not as bad as it seems — divide up the four processes of nitrogen fixation, ammonification, nitrification and denitrification and learn them separately. Then before you know it, you'll have learnt the whole cycle. Easy peesy.

Energy Transfer Through an Ecosystem

Energy is lost along food chains (how careless, they should put it in a safe place if you ask me).

Not All Energy gets Transferred to the Next Trophic Level

1) **Not all** the energy (e.g. from sunlight or food) that's available to the organisms in a trophic level is **transferred** to the **next** trophic level — around **90%** of the **total available energy** is **lost** in various ways.
2) Some of the available energy (**60%**) is **never taken in** by the organisms in the first place. For example:
   - Plants **can't use** all the light energy that reaches their **leaves**, e.g. some is the **wrong wavelength**, some is **reflected**, and some **passes straight through** the leaves.
   - Some sunlight can't be used because it hits parts of the plant that **can't photosynthesise**, e.g. the bark of a tree.
   - Some **parts** of food, e.g. **roots** or **bones**, **aren't eaten** by organisms so the energy isn't taken in.
   - Some parts of food are **indigestible** so **pass through** organisms and come out as **waste**, e.g. **faeces**. 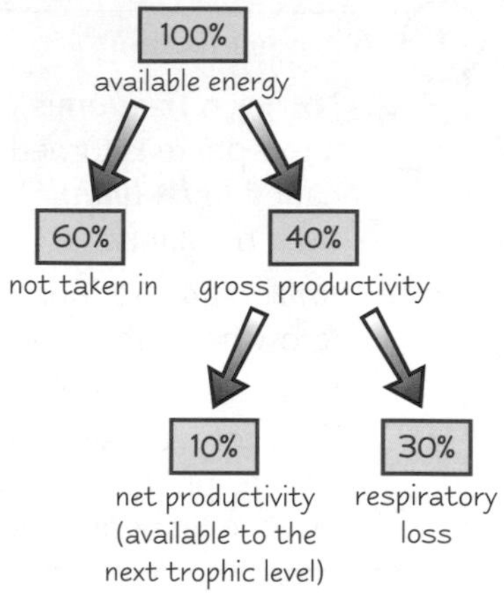
3) The rest of the available energy (**40%**) is **taken in** (**absorbed**) — this is called the **gross productivity**. But not all of this is available to the next trophic level either.
   - **30%** of the **total energy** available (75% of the gross productivity) is **lost to the environment** when organisms use energy produced from **respiration** for **movement** or body **heat**. This is called **respiratory loss**.
   - **10%** of the **total energy** available (25% of the gross productivity) becomes **biomass** (e.g. it's **stored** or used for **growth**) — this is called the **net productivity**.
4) **Net productivity** is the amount of energy that's **available** to the **next trophic level**. Here's how it's **calculated**:

100%
available energy
60%
not taken in
40%
gross productivity
10%
net productivity (available to the next trophic level)
30%
respiratory loss

net productivity = gross productivity – respiratory loss

EXAMPLE: The rabbits in an ecosystem receive **20 000 kJm^{-2} yr^{-1}** of energy, but don't take in **12 000 kJm^{-2} yr^{-1}** of it, so their gross productivity is **8000 kJm^{-2} yr^{-1}** (20 000 – 12 000). They lose **6000 kJm^{-2} yr^{-1}** using energy from **respiration**. You can use this to **calculate** the **net productivity** of the rabbits:

net productivity = 8000 – 6000
= 2000 kJm^{-2} yr^{-1}

5) You might be asked to **calculate** how **efficient energy transfer** from one trophic level to another is:

The rabbits receive **20 000 kJm^{-2} yr^{-1}**, and their **net productivity** is **2000 kJm^{-2} yr^{-1}**. So the **percentage efficiency** of **energy transfer** is: **(2000 ÷ 20 000) × 100 = 10%**

Energy Transfer Between Trophic Levels can be Measured

1) To **measure** the **energy transfer** between two trophic levels you need to **calculate** the **difference** between the amount of **energy** in each level (the net productivity of each level).
2) There are a couple of ways to **calculate** the **amount of energy** in a trophic level:

   1) You can **directly measure** the amount of **energy** (in **joules**) in the organisms by **burning** them in a **calorimeter**. The amount of **heat given off** tells you **how much** energy is in them.
   2) You can **indirectly measure** the amount of **energy** in the organisms by measuring their **dry mass** (their **biomass**). Biomass is created **using energy**, so it's an **indicator** of how much energy an organism **contains**.

3) First you calculate the amount of energy or biomass in a **sample** of the organisms, e.g. a 1 m^2 area of **wheat** or a single **mouse** that feeds on the wheat.
4) Then you **multiply** the results from the **sample** by the **size** of the **total population** (e.g. a 10 000 m^2 **field** of wheat or the **number** of mice in the population) to give the **total** amount of energy in the organisms at that **trophic level**.
5) The **difference** in **energy** between the trophic levels is the amount of energy **transferred**.
6) There are **problems** with this method though. For example, the consumers (mice) might have **taken in energy** from sources **other than** the producer measured (wheat). This means the difference between the two figures calculated **wouldn't** be an **accurate** estimate of the energy transferred between **only those two** organisms. For an **accurate estimate** you'd need to include **all** the individual organisms at each trophic level.

Energy Transfer Through an Ecosystem

Human Activities can Increase the Transfer of Energy Through an Ecosystem

Some **farming methods increase productivity** by **increasing** the **transfer** of **energy** through an **ecosystem**:

1) **Herbicides** kill **weeds** that **compete** with agricultural crops for **energy**. Reducing competition means crops receive **more energy**, so they grow **faster** and become **larger**, **increasing** productivity.
2) **Fungicides** kill **fungal infections** that **damage** agricultural crops. The crops **use more** energy for **growth** and **less** for fighting infection, so they grow **faster** and become **larger**, **increasing** productivity.
3) **Insecticides** kill **insect** pests that **eat** and **damage** crops. Killing insect pests means **less biomass** is **lost** from crops, so they grow to be **larger**, which means productivity is **greater**.
4) **Natural predators** introduced to the ecosystem **eat** the pest species, e.g. ladybirds eat greenfly. This means the crops lose **less energy** and **biomass**, **increasing** productivity.
5) **Fertilisers** are chemicals that provide crops with **minerals** needed **for growth**, e.g. **nitrates**. Crops **use up** minerals in the soil as they **grow**, so their growth is **limited** when there **aren't enough** minerals. Adding fertiliser **replaces** the lost minerals, so **more energy** from the ecosystem can be used to grow, **increasing** the **efficiency** of energy conversion.
6) Rearing livestock **intensively** involves **controlling** the **conditions** they live in, so **more** of their **energy** is used for **growth** and **less** is used for **other activities** — the **efficiency** of energy conversion is increased so **more biomass** is produced and productivity is **increased**. Here are a couple of **examples**:

   1) Animals may be kept in **warm**, **indoor** pens where their **movement** is **restricted**. **Less energy** is **wasted** keeping **warm** and **moving around**.
   2) Animals may be given **feed** that's **higher in energy** than their natural food. This **increases** the **energy input**, so **more energy** is available for **growth**.

   The benefits are that **more food** can be produced in a **shorter** space of time, often at **lower cost**. However, enhancing productivity by intensive rearing raises **ethical issues**. For example, some people think the **conditions** intensively reared animals are kept in cause the animals **pain**, **distress** or restricts their **natural behaviour**, so it **shouldn't be done**.

Practice Questions

Q1 What is the equation for net productivity?

Q2 What do you need to calculate to find the energy transfer between two trophic levels?

Q3 Give one example of how farmers increase the productivity of crops.

Q4 Give one example of how farmers increase the productivity of animals.

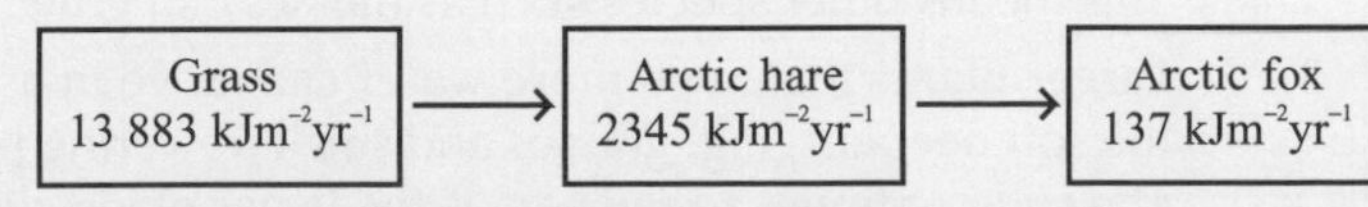

Exam Question

Q1 The diagram above shows the net productivity of different trophic levels in a food chain.

a) Explain why the net productivity of the Arctic hare is less than the net productivity of the grass. [4 marks]

b) Calculate the percentage efficiency of energy transfer from the Arctic hare to the Arctic fox. [2 marks]

I'm suffering from energy loss after those two pages...

Crikey, so farming's not just about getting up early to feed the chooks then — farmers are manipulating the transfer of energy to produce as much food as they can. And it's really important to remember that this transfer of energy isn't 100% efficient — most gets lost along the way so the next organisms don't get all the energy. Interesting, ve-ry interesting....

Succession

You've read that ecosystems are dynamic systems — they change over time. So now you need to know how they change.

Succession is the *Process of Ecosystem Change*

Succession is the process by which an **ecosystem changes** over **time**. The **biotic conditions** (e.g. **plant** and **animal communities**) change as the **abiotic conditions** change (e.g. **water** availability). There are **two** types of succession:

1) **Primary succession** — this happens on land that's been **newly formed** or **exposed**, e.g. where a **volcano** has erupted to form a **new rock surface**, or where **sea level** has **dropped** exposing a new area of land. There's **no soil** or **organic material** to start with, e.g. just bare rock.
2) **Secondary succession** — this happens on land that's been **cleared** of all the **plants**, but where the **soil remains**, e.g. after a **forest fire** or where a forest has been **cut down by humans**.

Succession Occurs in *Stages called Seral Stages*

1) **Primary succession** starts when species **colonise** a new land surface. **Seeds** and **spores** are blown in by the **wind** and begin to **grow**. The **first species** to colonise the area are called **pioneer species** — this is the **first seral stage**.
    - The **abiotic conditions** are **hostile** (**harsh**), e.g. there's no soil to **retain water**. Only pioneer species **grow** because they're **specialised** to cope with the harsh conditions, e.g. **marram grass** can grow on sand dunes near the sea because it has **deep roots** to get water and can **tolerate** the salty environment.
    - The pioneer species **change** the **abiotic conditions** — they **die** and **microorganisms decompose** the dead **organic material** (**humus**). This forms a **basic soil**.
    - This makes conditions **less hostile**, e.g. the basic soil helps to **retain water**, which means **new organisms** can move in and grow. These then die and are decomposed, adding **more** organic material, making the soil **deeper** and **richer in minerals**. This means **larger plants** like **shrubs** can start to grow in the deeper soil, which retains **even more** water. As **more plants** move in they create **more habitats**, so **more animals** move in.
2) **Secondary succession** happens in the **same way**, but because there's already a **soil layer** succession starts at a **later seral stage** — the pioneer species in secondary succession are **larger plants**, e.g. shrubs.
3) At each stage, **different** plants and animals that are **better adapted** for the improved conditions move in, **out-compete** the plants and animals that are already there, and become the **dominant species** in the ecosystem.
4) As succession goes on, the ecosystem becomes **more complex**. New species move in **alongside** existing species, which means the **species diversity** (the number of **different species** and the **abundance** of each species) **increases**.
5) The amount of **biomass** also **increases** because plants at later stages are **larger** and **more dense**, e.g. **woody trees**.
6) The **final seral stage** is called the **climax community** — the ecosystem is supporting the **largest** and **most complex** community of plants and animals it can. It **won't change** much more — it's in a **steady state**.

This example shows primary succession on bare rock, but succession also happens on sand dunes, salt marshes and even on lakes.

Example of primary succession — bare rock to woodland

1) **Pioneer species colonise** the rocks. E.g. **lichens** grow **on** and **break down** rocks, **releasing minerals**.
2) The lichens **die** and are **decomposed** helping to form a **thin soil**, which thickens as more **organic material** is formed. This means other species such as **mosses** can **grow**.
3) **Larger plants** that need **more water** can move in as the soil **deepens**, e.g. **grasses** and **small flowering plants**. The soil **continues to deepen** as the larger plants die and are decomposed.
4) **Shrubs**, **ferns** and **small trees** begin to grow, **out-competing** the grasses and smaller plants to become the **dominant** species. **Diversity increases**.
5) Finally, the soil is **deep** and **rich** enough in **nutrients** to support **large trees**. These become the dominant species, and the **climax community** is formed.

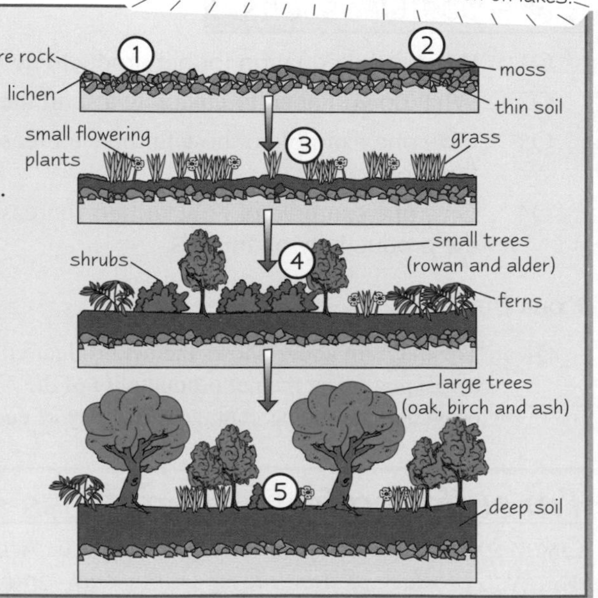

Succession

Different Ecosystems have Different Climax Communities

Which species make up the climax community depends on what the **climate's** like in an ecosystem. The climax community for a **particular** climate is called its **climatic climax**. For example:

> In a **temperate climate** there's **plenty** of **available water, mild temperatures** and not much **change** between the seasons. The climatic climax will contain **large trees** because they **can grow** in these conditions once **deep soils** have developed. In a **polar climate** there's **not much available water**, temperatures are **low** and there are **massive changes** between the seasons. Large trees **won't ever** be able to grow in these conditions, so the climatic climax contains only **herbs** or **shrubs**, but it's still the **climax community**.

Succession can be Prevented or Deflected

Human activities can **prevent succession**, stopping the normal climax community from **developing**. When succession is stopped **artificially** like this, the climax community is called a **plagioclimax**. **Deflected succession** is when succession is prevented by human activity, but the plagioclimax that develops is one that's **different** to any of the **natural seral stages** of the ecosystem — the path of succession has been **deflected** from its natural course.

For example:

1) A **regularly mown** grassy field **won't develop** woody plants, even if the climate of the ecosystem could support them.

Man had been given a mighty weapon with which they would tame the forces of nature

2) The **growing points** of the woody plants are **cut off** by the lawnmower, so larger plants **can't establish** themselves — only the grasses can **survive** being mowed, so the **climax community** is a **grassy field**.

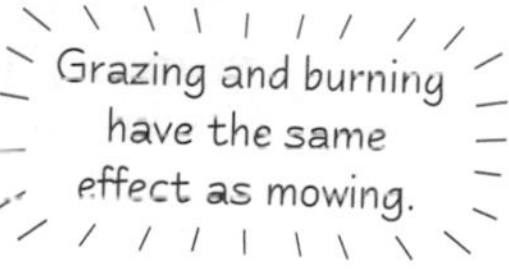

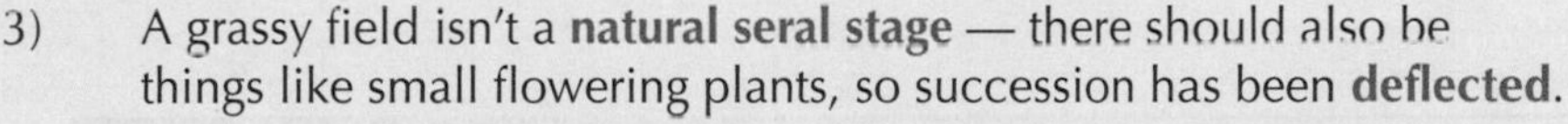

3) A grassy field isn't a **natural seral stage** — there should also be things like small flowering plants, so succession has been **deflected**.

Practice Questions

Q1 What is the difference between primary and secondary succession?

Q2 What is the name given to species that are the first to colonise an area during succession?

Q3 What is meant by a climax community?

Exam Question

Q1 A farmer has a field where he plants crops every year. When the crops are fully grown he removes them all and then ploughs the field (churns up all the plants and soil so the field is left as bare soil). The farmer has decided not to plant crops or plough the field for several years.

a) Describe, in terms of succession, what will happen in the field over time. [6 marks]

b) Explain why succession doesn't usually take place in the farmer's field. [2 marks]

Revision succession — bare brain to a woodland of knowledge...

When answering questions on succession, examiners are pretty keen on you using the right terminology — that means saying "pioneer species" instead of "the first plants to grow there". If you can manage that, then you can manage succession.

Investigating Ecosystems

Examiners aren't happy unless you're freezing to death in the rain in a field somewhere in the middle of nowhere. Still, it's better than being stuck in the classroom being bored to death learning about fieldwork techniques...

You need to be able to **Investigate Populations** of **Organisms**

Investigating **populations** of organisms involves looking at the **abundance** and **distribution** of **species** in a particular **area**.

1) **Abundance** — the **number of individuals** of **one species** in a **particular area.** The abundance of **mobile organisms** and **plants** can be estimated by simply counting the **number** of individuals in samples taken. **Percentage cover** can also be used to measure the abundance of plants — this is **how much** of the area you're investigating is **covered** by a species.
2) **Distribution** — this is **where** a particular species is within the **area you're investigating**.

You need to take a **Random Sample** from the **Area You're Investigating**

Most of the time it would be too **time-consuming** to measure the **number of individuals** and the **distribution** of every species in the **entire area** you're investigating, so instead you take **samples**:

1) **Choose** an **area** to **sample** — a **small** area **within** the area being investigated.
2) Samples should be **random** to **avoid bias**, e.g. if you were investigating a field you could pick random sample sites by dividing the field into a **grid** and using a **random number generator** to select **coordinates.**
3) Use an **appropriate technique** to take a sample of the population (see below and on the next page).
4) **Repeat** the process, taking as many samples as possible. This gives a more **reliable** estimate for the **whole area.**
5) The **number of individuals** for the **whole area** can then be **estimated** by taking an **average** of the data collected in each sample and **multiplying** it by the size of the whole area. The **percentage cover** for the whole area can be estimated by taking the average of all the samples.

Finally! 26 542 981 poppies. What do you mean I didn't need to count them all?

Frame Quadrats can be used to **Investigate Plant Populations**

1) A **frame quadrat** is a **square** frame divided into a **grid** of 100 **smaller squares** by strings attached across the frame.
2) They're **placed on the ground** at **random points** within the area you're investigating. This can be done by selecting **random coordinates** (see above).
3) The **number of individuals** of each species is recorded in **each quadrat.**
4) The **percentage cover** of a species can also be measured by counting how much of the quadrat is **covered** by the species — you count a square if it's **more than half-covered.** Percentage cover is a **quick** way to investigate populations and you **don't** have to **count** all the **individual** plants.
5) Frame quadrats are useful for **quickly** investigating areas with species that **fit** within a **small quadrat** — most frame quadrats are **1 m by 1 m.**
6) Areas with **larger plants** and **trees** need **very large** quadrats. Large quadrats **aren't** always in a frame — they can be marked out with a **tape measure.**

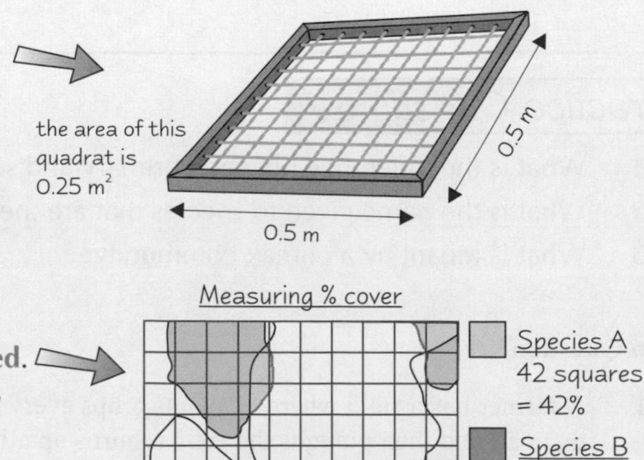

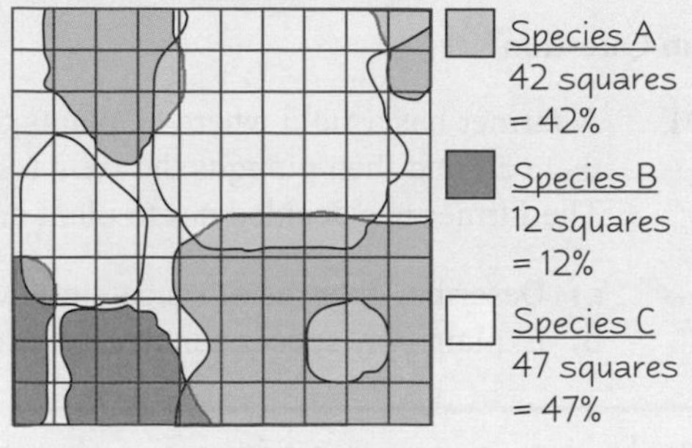

Investigating Ecosystems

Point Quadrats can also be used to **Investigate Plant Populations**

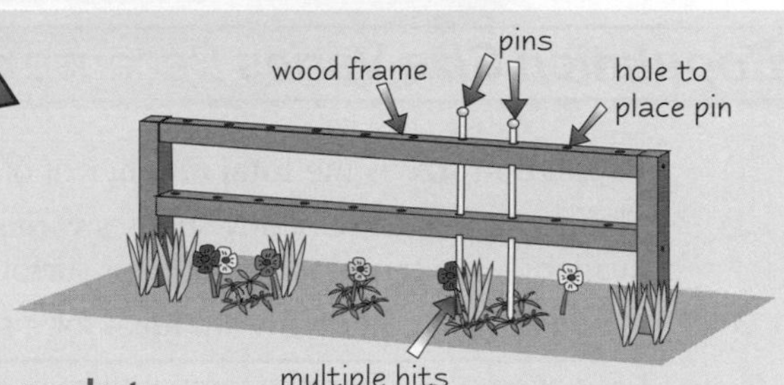

1) A **point quadrat** is a **horizontal bar** on **two legs** with a series of holes at set intervals along its length.
2) Point quadrats are **placed on the ground** at **random points** within the area you're investigating.
3) **Pins** are dropped through the holes in the frame and **every plant** that each pin **touches** is **recorded**. If a pin touches several **overlapping** plants, **all** of them are recorded.
4) The **number of individuals** of each species is recorded in **each quadrat**.

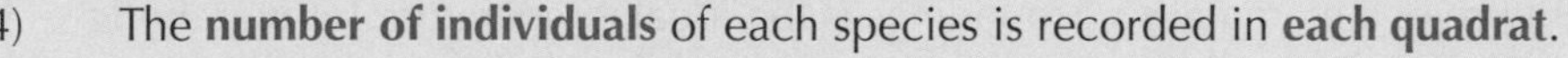

5) The **percentage cover** of a species can also be measured by calculating the **number of times** a pin has touched a species as a **percentage** of the **total number** of pins dropped.

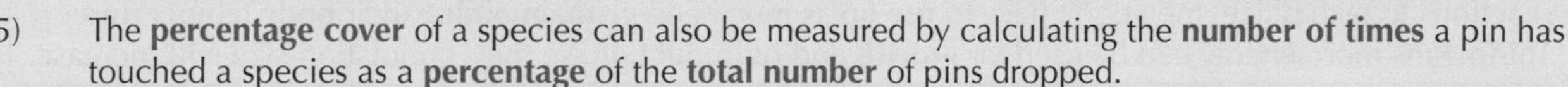

6) Point quadrats are especially useful in areas where there's lots of **dense vegetation** close to the ground.

Transects are used to **Investigate** the **Distribution** of **Plant Populations**

You can use **lines** called **transects** to help find out how plants are **distributed across** an area, e.g. how species **change** from a hedge towards the middle of a field. You need to know about **three** types of transect:

1) **Line transects** — a **tape measure** is placed **along** the transect and the species that **touch** the tape measure are **recorded**.

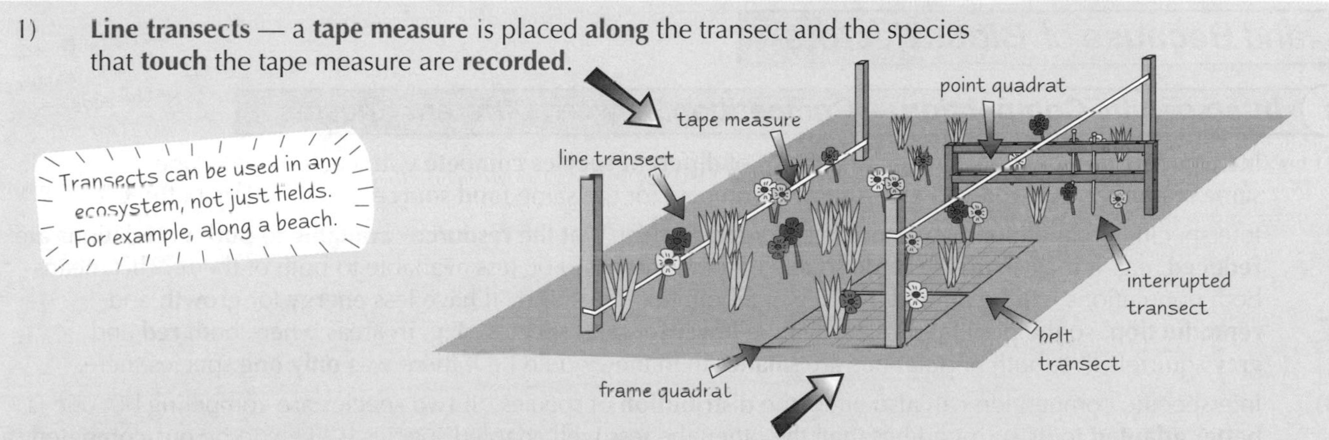

2) **Belt transects** — data is collected along the transect using **frame quadrats** placed **next to** each other.
3) **Interrupted transects** — instead of investigating the **whole transect** of either a line or a belt, you can take **measurements** at **intervals**. E.g. by placing **point quadrats** at **right angles** to the direction of the transect at **set intervals** along its length, such as **every 2 m**.

Practice Questions

Q1 Define abundance.

Q2 What does percentage cover show?

Q3 Explain why samples of a population are taken.

Q4 Briefly describe how belt transects are different from line transects.

Exam Question

Q1 A student wants to sample a population of daffodils in a field.

a) How could she avoid bias in her investigation? [1 mark]

b) Describe how she could investigate the percentage cover of daffodils in the field using frame quadrats. [3 marks]

What did the quadrat say to the policeman — I've been framed...

If you want to know what it's really like doing these investigations then read these pages outside in the pouring rain. Doing it while you're tucked up in a nice warm, dry exam hall won't seem so bad after that, take my word for it.

Factors Affecting Population Size

Uh-oh, anyone who loves cute little bunny-wunnys look away now — these pages are about how the population sizes of organisms fluctuate and the reasons why. One of the reasons, I'm sad to say, is because the little rabbits get eaten.

Population Size Varies Because of Abiotic Factors...

Remember — abiotic factors are the non-living features of the ecosystem.

1) **Population size** is the **total number** of organisms of **one species** in a **habitat**.
2) The **population size** of any species **varies** because of **abiotic** factors, e.g. the amount of **light**, **water** or **space** available, the **temperature** of their surroundings or the **chemical composition** of their surroundings.
3) When abiotic conditions are **ideal** for a species, organisms can **grow fast** and **reproduce successfully**.

   For example, when the temperature of a mammal's surroundings is the ideal temperature for **metabolic reactions** to take place, they don't have to **use up** as much energy **maintaining** their **body temperature**. This means more energy can be used for **growth** and **reproduction**, so their population size will **increase**.

4) When abiotic conditions **aren't ideal** for a species, organisms **can't** grow as **fast** or reproduce as **successfully**.

   For example, when the temperature of a mammal's surroundings is significantly **lower** or **higher** than their **optimum** body temperature, they have to **use** a lot of **energy** to maintain the right **body temperature**. This means less energy will be available for **growth** and **reproduction**, so their population size will **decrease**.

...and Because of Biotic Factors

Biotic factors are the living features of the ecosystem.

1 Interspecific Competition — Competition Between Different Species

1) Interspecific competition is when organisms of **different species compete** with each other for the **same resources**, e.g. **red** and **grey** squirrels compete for the same **food sources** and **habitats** in the **UK**.
2) Interspecific competition between two species can mean that the **resources available** to **both** populations are **reduced**, e.g. if they share the **same** source of food, there will be **less** available to both of them. This means both populations will be **limited** by a lower amount of food. They'll have less **energy** for **growth** and **reproduction**, so the population sizes will be **lower** for both species. E.g. in areas where both **red** and **grey** squirrels live, both populations are **smaller** than they would be if there was **only one** species there.
3) Interspecific competition can also affect the **distribution** of species. If **two** species are competing but one is **better adapted** to its surroundings than the other, the less well adapted species is likely to be **out-competed** — it **won't** be able to **exist** alongside the better adapted species. E.g. since the introduction of the **grey squirrel** to the UK, the native **red squirrel** has **disappeared** from large areas. The grey squirrel has a better chance of **survival** because it's **larger** and can store **more fat** over winter.

Plants compete for things like minerals and light.

2 Intraspecific Competition — Competition Within a Species

Intraspecific competition is when organisms of the **same species compete** with each other for the **same resources**.

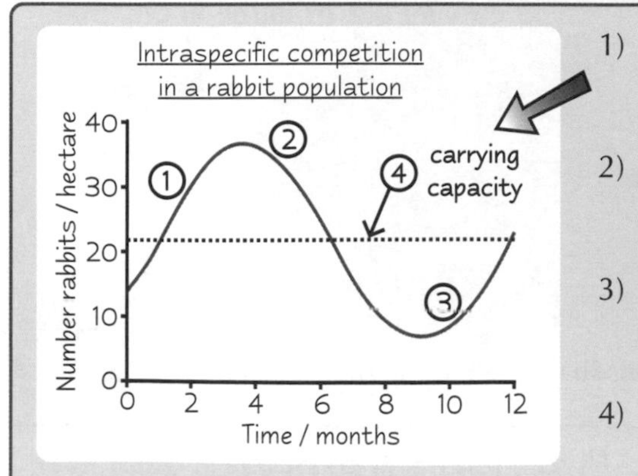

1) The **population** of a species (e.g. rabbits) **increases** when resources are **plentiful**. As the population increases, there'll be **more** organisms competing for the **same amount** of **space** and **food**.
2) Eventually, resources such as food and space become **limiting** — there **isn't enough** for all the organisms. The population then begins to **decline**.
3) A **smaller** population then means that there's **less competition** for space and food, which is **better** for **growth** and **reproduction** — so the population starts to **grow** again.
4) The **maximum stable population size** of a species that an ecosystem can **support** is called the **carrying capacity**.

Factors Affecting Population Size

3) Predation — Predator and Prey Population Sizes are Linked

Predation is where an organism (the predator) kills and eats another organism (the prey), e.g. lions kill and eat (**predate** on) buffalo. The **population sizes** of predators and prey are **interlinked** — as the population of one **changes**, it **causes** the other population to **change**:

1) As the **prey** population **increases**, there's **more food** for predators, so the **predator** population **grows**. E.g. in the graph on the right the **lynx** population **grows** after the **snowshoe hare** population has **increased** because there's **more food** available.
2) As the **predator** population **increases**, **more prey** is **eaten** so the **prey** population then begins to **fall**. E.g. **greater numbers** of lynx eat lots of snowshoe hares, so their population **falls**.
3) This means there's **less food** for the **predators**, so their population **decreases**, and so on. E.g. **reduced** snowshoe hare numbers means there's **less food** for the lynx, so their population **falls**.

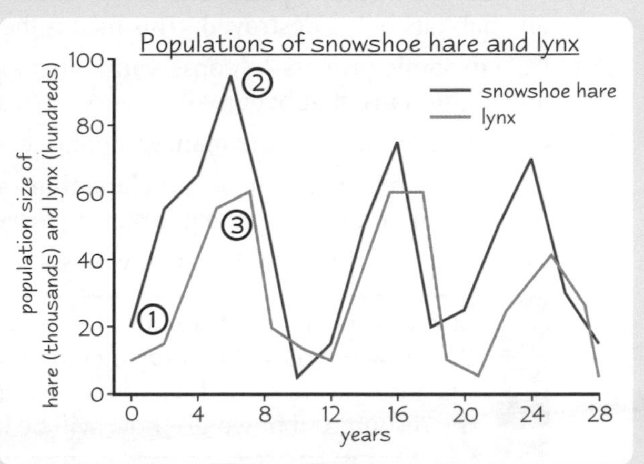

Predator-prey relationships are usually more **complicated** than this though because there are **other factors** involved, like availability of **food** for the **prey**. E.g. it's thought that the population of snowshoe hare initially begins to **decline** because there's **too many** of them for the amount of **food available**. This is then **accelerated** by **predation** from the lynx.

Limiting Factors Stop the Population Size of a Species Increasing

1) Limiting factors can be **abiotic**, e.g. the amount of **shelter** in an ecosystem **limits** the population size of a species because there's only enough shelter for a **certain number** of individuals.
2) Limiting factors can also be **biotic**, e.g. **interspecific competition limits** the population size of a species because the amount of **resources** available to a species is **reduced**.

Practice Questions

Q1 What is interspecific competition?

Q2 What will be the effect of interspecific competition on the population size of a species?

Q3 Define intraspecific competition.

Q4 What does 'carrying capacity' mean?

Q5 What is a limiting factor?

Exam Question

Q1 The graph on the right shows the population size of a predator species and a prey species over a period of 30 years.

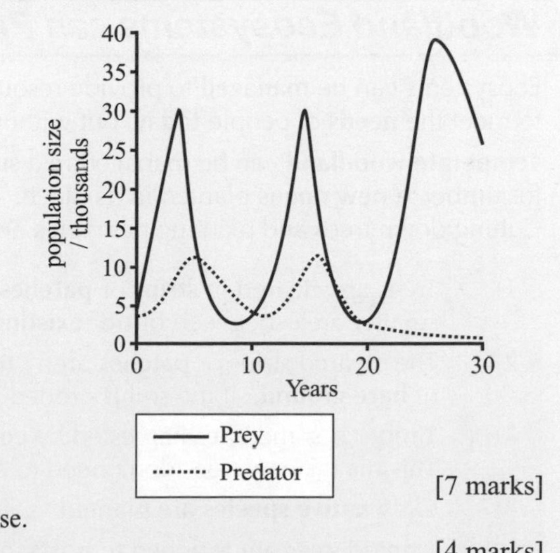

a) Using the graph, describe and explain how the population sizes of the predator and prey species vary over the first 20 years. [7 marks]

b) The numbers of species B declined after year 20 because of a disease. Describe and explain what happened to the population of species A. [4 marks]

Predator-prey relationships — they don't usually last very long...

You'd think they could have come up with names a little more different than inter- and intraspecific competition. I always remember it as int-er means diff-er-ent species. The factors that affect population size are divided up nicely for you here — abiotic factors, competition and predation — just like predators like to nicely divide up their prey into bitesize chunks.

Conservation of Ecosystems

It's important that ecosytems are conserved so the resources we use from them to make lots of nice things don't run out.

We Need to Conserve Ecosystems

1) **Conservation** is the **protection** and **management** of **ecosystems** so that the **natural resources** in them can be **used** without them **running out**. E.g. using rainforests for timber without any species becoming **extinct** and without any habitats being **destroyed**. This means the natural resources will still be available for **future generations.**
2) It's a **dynamic process** — conservation methods need to be **adapted** to the **constant changes** (caused **naturally** and by **humans**) that occur within ecosystems.
3) Conservation involves the **management** of ecosystems — controlling how **resources** are **used** and **replaced**.
4) Conservation can also involve **reclamation** — **restoring ecosystems** that have been **damaged** or **destroyed** so they can be **used again**, e.g. restoring **forests** that have been **cut down** so they can be used again.
5) Conservation is **important** for many reasons:

Economic: **Ecosystems** provide **resources** for lots of things that **humans need**, e.g. **rainforests** contain species that provide things like **drugs**, **clothes** and **food**. These resources are **economically important** because they're **traded** on a **local** and **global** scale. If the ecosystems **aren't** conserved, the resources that we use now will be **lost**, so there will be **less trade** in the future.

Social: Many ecosystems bring **joy** to lots of people because they're **attractive** to **look at** and people **use** them for **activities**, e.g. birdwatching and walking. The species and habitats in the ecosystems may be **lost** if they **aren't** conserved, so **future generations** won't be able to use and enjoy them.

Ethical:
1) Some people think we should conserve ecosystems simply because it's the **right thing to do**, e.g. most people think organisms have a **right to exist**, so they shouldn't become extinct as a result of **human activity**.
2) Some people think we have a **moral responsibility** to conserve ecosystems for **future generations**, so they can enjoy and use them.

Cast your mind back to AS biology — the reasons for conservation are similar to the reasons for conserving biodiversity.

6) **Preservation** is different from conservation — it's the **protection** of ecosystems so they're kept **exactly as they are**. Nothing is **removed** from a preserved ecosystem and they're only **used** for activities that **don't damage** them. For example, **Antarctica** is a preserved ecosystem because it's protected from **exploitation** by humans — it's only used for **limited tourism** and **scientific research**, not **mining** or other **industrial** activities.

Woodland Ecosystems *can* Provide Resources *in a* Sustainable Way

Ecosystems can be **managed** to provide resources in a way that's **sustainable** — this means enough resources are taken to meet the **needs** of people **today**, but without **reducing the ability** of people in the **future** to meet their own needs.

Temperate woodland can be managed in a **sustainable way** — for every tree that's **cut down** for timber, a **new one** is planted in its place. The woodland should never become **depleted**. Cutting down trees and planting new ones needs to be done **carefully** to be **successful**:

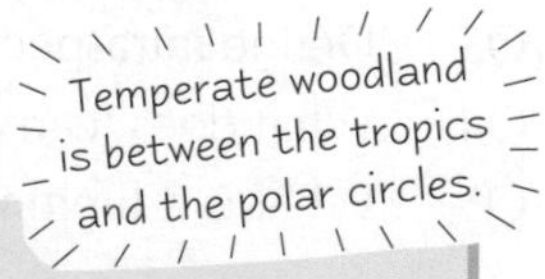

1) Trees are cleared in **strips** or **patches** — woodland grows back **more quickly** in smaller areas between bits of **existing woodland** than it does in larger, **open areas.**
2) The cleared strips or patches aren't **too large** or **exposed** — lots of **soil erosion** can occur on large areas of **bare ground**. If the soil is eroded, newly planted trees **won't** be able to **grow**.
3) Timber is sometimes harvested by **coppicing** — **cutting** down trees in a way that lets them **grow back**. This means new trees don't need to be planted.
4) Only **native species** are planted — they grow most **successfully** because they're **adapted** to the climate.
5) Planted trees are attached to **posts** to provide **support**, and are grown in **plastic tubes** to stop them being **eaten** by grazing animals — this makes it **more likely** the trees will **survive** to become mature adults.
6) Trees **aren't** planted too **close together** — this means the trees aren't **competing** with each other for **space** or **resources**, so they're more likely to **survive**.

Conservation of Ecosystems

Human Activities Affect Ecosystems like the Galapagos Islands

Humans often need to **conserve** or **preserve** ecosystems because our **activities** have **badly affected** them, e.g. large areas of the **Amazon rainforest** have been **cleared** without being **replaced**, **destroying** the ecosystem.

Human activities have had a negative effect on the **Galapagos Islands**, a small group of islands in the **Pacific Ocean** about 1000 km off the coast of South America. Many species of animals and plants have evolved there that **can't** be found **anywhere else**, e.g. the **Galapagos giant tortoise** and the **Galapagos sea lion**. Here are some examples of how the **animal** and **plant populations** there have been affected by human activity:

1) **Explorers** and **sailors** that visited the Galapagos Islands in the **19th century** directly affected the populations of some animals by **eating them**. For example, a type of **giant tortoise** found on **Floreana Island** was hunted to **extinction** for food.
2) **Non-native animals introduced** to the islands **eat** some native species. This has caused a decrease in the populations of native species. For example, non-native **dogs, cats** and **black rats** eat young **giant tortoises** and **Galapagos land iguanas. Pigs** also destroy the nests of the iguanas and **eat their eggs. Goats** have eaten much of the **plant life** on some of the islands.
3) **Non-native plants** have also been introduced to the islands. These **compete** with native plant species, causing a decrease in their populations. For example, **quinine trees** are **taller** than some native plants — they **block out light** to the native plants, which then **struggle** to **survive**.
4) **Fishing** has caused a decrease in the populations of some of the **sea life** around the Galapagos Islands. For example, the populations of **sea cucumbers** and **hammerhead sharks** have been reduced because of **overfishing. Galapagos green turtle** numbers have also been reduced by overfishing and they're also killed **accidentally** when they're caught in **fishing nets**. They're now an **endangered species**.
5) A recent increase in **tourism** (from **41 000** tourists in **1991** to around **160 000** in **2008**) has led to an increase in **development** on the islands. For example, the **airport** on Baltra island has been redeveloped to receive more tourists. This causes **damage** to the ecosystems as **more land** is **cleared** and **pollution** is **increased**.
6) The **population** on the islands has also **increased** due to the increased **opportunities** from tourism. This could lead to further **development** and so more **damage** to the ecosystems.

Darwin (the sea lion) worried he was about to be affected by human activity.

Practice Questions

Q1 Why does conservation need to be dynamic?

Q2 What is meant by reclamation?

Q3 How is preservation different from conservation?

Q4 What does managing an ecosystem in a sustainable way mean?

Q5 Give one way that temperate woodlands are managed to make sure newly planted trees grow.

Exam Questions

Q1 Explain why conservation is important for economic, social and ethical reasons. [3 marks]

Q2 Explain how the following human activities have affected specific native animal or plant populations on the Galapagos Islands.

a) Introduction of non-native animal species. [2 marks]

b) Introduction of non-native plant species. [2 marks]

c) Fishing. [2 marks]

If I can sustain this revision it'll be a miracle...

Never mind ecosystems, I'm more interested in preserving my sanity after all this hard work. I know it doesn't seem all that sciencey, but you can still study biology without a lab coat, some Petri dishes and a rack of test tubes. Sustainability's a funny one to get your head around, but luckily you just need to know about how it applies to temperate woodlands.

Plant Responses

You might not think that plants do much, but they respond to stimuli just like us. OK, not just like us (I can't picture a daisy boogying to cheesy music), but their responses are important all the same. So important there's four pages of em'...

Plants Need to Respond to Stimuli Too

1) Plants, like animals, **increase** their chances of **survival** by **responding** to changes in their **environment**, e.g:

- They sense the direction of **light** and **grow** towards it to **maximise** light absorption for **photosynthesis**.
- They can sense **gravity**, so their roots and shoots **grow** in the **right direction**.
- **Climbing** plants have a sense of **touch**, so they can find things to climb and **reach** the **sunlight**.

2) Plants are more likely to survive if they **respond** to the presence of **predators** to **avoid being eaten**, e.g. some plants produce **toxic substances**:

White clover is a plant that can produce substances that are **toxic** to **cattle**. Cattle start to **eat** lots of white clover when fields are **overgrazed** — the white clover **responds** by **producing toxins**, to **avoid** being **eaten**.

3) Plants are more likely to survive if they **respond** to **abiotic stress** — anything **harmful** that's **natural** but **non-living**, like a **drought**. E.g. some plants respond to **extreme cold** by **producing** their own form of **antifreeze**:

Carrots produce **antifreeze proteins** at low temperatures — the proteins **bind** to **ice crystals** and **lower** the **temperature** that water **freezes** at, **stopping** more ice crystals from **growing**.

A Tropism is a Plant's Growth Response to an External Stimulus

1) A **tropism** is the **response** of a plant to a **directional stimulus** (a stimulus coming from a particular direction).
2) Plants respond to directional stimuli by **regulating** their **growth**.
3) A **positive tropism** is growth **towards** the stimulus.
4) A **negative tropism** is growth **away** from the stimulus.
5) An example of a tropism is **phototropism** — the growth of a plant in response to **light**:

- **Shoots** are **positively phototropic** and grow **towards** light.
- **Roots** are **negatively phototropic** and grow **away** from light.

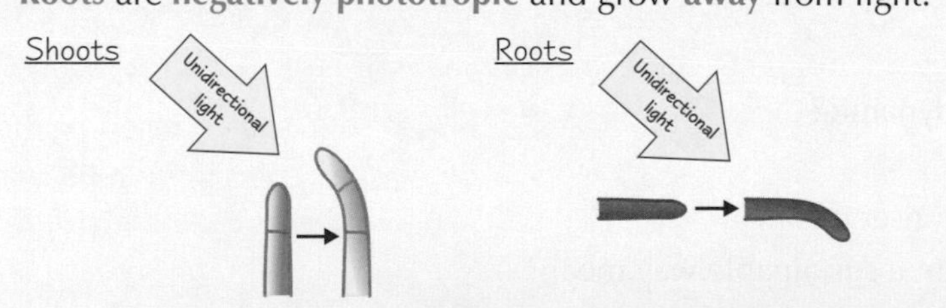

The men's gymnastics team were positively phototropic.

Responses are Brought About by Growth Hormones

Growth hormones are also called growth substances.

1) Plants **respond** to stimuli using **growth hormones** — these are chemicals that **speed up** or **slow down** plant **growth**.
2) Growth hormones are **produced** in the **growing regions** of the plant (e.g. shoot tips, leaves) and they **move** to where they're needed in the **other parts** of the plant.
3) A growth hormone called **gibberellin** stimulates **seed germination**, **stem elongation**, **side shoot formation** and **flowering**.
4) Growth hormones called **auxins** stimulate the **growth** of shoots by **cell elongation** — this is where **cell walls** become **loose** and **stretchy**, so the cells get **longer**.
5) **High** concentrations of auxins **inhibit growth** in **roots** though.

Plant Responses

The Uneven Distribution of Auxins Causes Uneven Growth

1) **Auxins** are produced in the **tips** of **shoots** in flowering plants (called **apical buds**).
2) **Indoleacetic acid** (**IAA**) is an important **auxin** that's involved in **phototropism**.
3) Auxins (including IAA) are **moved** around the plant to **control tropisms** — they move by **diffusion** and **active transport** over short distances, and via the **phloem** over longer distances.
4) This results in **different parts** of the plants having **different amounts** of auxins. The **uneven distribution** of auxins means there's **uneven growth** of the plant, e.g:

Phototropism — auxins move to the more **shaded** parts of the **shoots** and **roots**, so there's uneven growth.

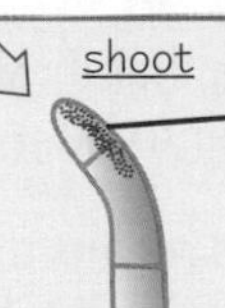

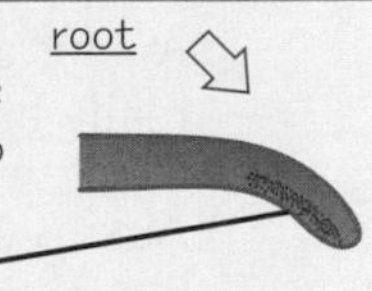

Auxins are Involved in Apical Dominance

1) Auxins **stimulate** the **growth** of the **apical bud** and **inhibit** the **growth** of **side shoots**. This is called **apical dominance** — the apical bud is **dominant** over the side shoots.

2) Apical dominance prevents side shoots from growing — this **saves energy** and prevents side shoots from the same plant **competing** with the shoot tip for light.
3) Because energy **isn't** being used to grow side shoots, apical dominance allows a **plant** in an area where there are **loads of other plants** to **grow tall very fast**, past the smaller plants, to **reach** the **sunlight**.
4) If you **remove** the apical bud then the plant **won't produce auxins**, so the **side shoots** will **start growing** by **cell division** and **cell elongation**.

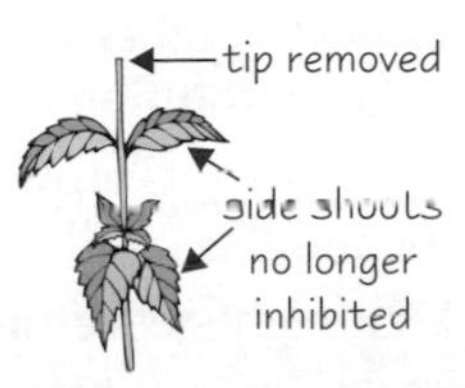

5) Auxins become **less concentrated** as they **move away** from the apical bud to the rest of the plant. If a plant grows **very tall**, the bottom of the plant will have a **low auxin concentration** so side shoots will start to grow near the bottom.

The Role of Auxins in Apical Dominance Can be Investigated Experimentally

Here's an example of how you do it:

1) Plant **30 plants** (e.g. **pea plants**) that are a **similar age**, **height** and **weight** in pots.
2) **Count** and **record** the number of **side shoots** growing from the main stem of **each plant**.
3) For **10 plants**, **remove** the **tip** of the **shoot** and apply a **paste containing auxins** to the **top** of the **stem**.
4) For another 10 plants, remove the tip of the shoot and apply a **paste without auxins** to the top of the stem.
5) Leave the final 10 plants as they are — these are your untouched **controls**.
6) Remember, you **always** need to have controls (e.g. without the hormone, untouched) for **comparison** — so you know the **effect** you see is **likely** to be due to the **hormone** and **not any other factor**.
7) Let each group **grow** for about **six days**. You need to keep all the plants in the **same conditions** — the same **light intensity**, **water**, etc. This makes sure any **variables** that may affect your results are **controlled**, which makes your experiment **more reliable**.
8) After six days, **count** the number of **side shoots** growing from the main stem of **each** of your **plants**.
9) You might get **results** a bit like these:

	plants left untreated (control group)	tips removed, paste with auxins applied	tips removed, paste without auxins applied
average no. of side shoots per plant at start of experiment	4	4	4
average no. of side shoots per plant at end of experiment	5	5	9

10) The results in the **table** show that **removing** the **tips** of shoots caused **extra side shoots** to **grow**, but removing tips **and** applying **auxins** **reduced the number** of extra side shoots.
11) The results suggest auxins **inhibit** the **growth** of side shoots — suggesting that auxins are involved in **apical dominance**.

Plant Responses

Gibberellins and Auxins can Work Together

1) **Gibberellins** are produced in **young leaves** and in **seeds**.
2) They stimulate seed germination, stem elongation, side shoot formation and flowering.
3) Gibberellins **stimulate** the **stems** of plants to **grow** by **stem elongation** — this helps plants to grow **very tall**. If a **dwarf variety** of a plant is treated with gibberellin, it will grow to the **same height** as the **tall variety**.
4) Unlike auxins, gibberellins **don't inhibit** plant growth in any way.
5) **Auxins** and **gibberellins** sometimes **work together** to affect plant growth:

Auxins and gibberellins are often **synergistic** — this means that they **work together** to have a really **big effect**. E.g. auxins and gibberellins work together to help plants grow **very tall**.

Auxins and gibberellins are sometimes **antagonistic** — this means they **oppose** each other's actions. E.g. **gibberellins stimulate** the growth of **side shoots** but **auxins inhibit** the growth of side shoots.

The Role of Gibberellins in Stem Elongation Can be Investigated

Here's an example of how you do it:

1) Plant **40 plants** (e.g. **dwarf pea plants**) that are a **similar age**, **height** and **mass** in pots.
2) **Leave 20** plants as they are to grow, **watering** them **all** in the **same way** and keeping them **all** in the **same conditions** — these are your **controls**.
3) **Leave** the **other 20 plants** to grow in the **same conditions**, **except** water them with a **dilute solution** of **gibberellin** (e.g. **100 μg/ml** gibberellin).
4) Let the plants grow for about **28 days** and **measure** the **lengths** of all the **stems once each week**.
5) You might get **results** a bit like these: ⇨
6) The results in the **table** show that stems **grow more** when watered with a dilute solution of **gibberellin**.
7) The results suggest **gibberellin stimulates stem elongation**.
8) You might have to **calculate** the **rate of growth** of the plants in your exam, e.g:
   - In **28 days** the plants **watered normally** grew an **average** of **9 cm** (23 cm – 14 cm), so they grew at an average **rate** of 9 ÷ 28 = **0.32 cm/day**.
   - In **28 days** the plants **watered with gibberellin** grew an **average** of **32 cm** (46 cm – 14 cm), so they grew at an average **rate** of 32 ÷ 28 = **1.14 cm/day**.

time / days	average stem length / cm	
	plants watered normally	plants watered with gibberellin
0	14	14
7	15	17
14	18	27
21	19	38
28	23	46

Hormones are Involved in Leaf Loss in Deciduous Plants

1) **Deciduous plants** are plants that **lose** their **leaves** in **winter**.
2) Losing their leaves helps plants to **conserve water** (lost from leaves) during the cold part of the year, when it might be **difficult** to **absorb water** from the **soil** (the soil water may be **frozen**), and when there's **less light** for **photosynthesis**.
3) Leaf loss is **triggered** by the **shortening day length** in the autumn.
4) Leaf loss is **controlled** by **hormones**:

The technical term for leaf loss is abscission.

- **Auxins inhibit** leaf loss — auxins are produced by **young leaves**. As the leaf gets **older**, **less auxin** is produced, leading to **leaf loss**.
- **Ethene stimulates** leaf loss — ethene is produced by **ageing leaves**. As the leaves get **older**, **more ethene** is produced. A **layer** of **cells** (called the **abscission layer**) develops at the **bottom** of the **leaf stalk** (where the leaf joins the stem). The abscission layer **separates** the leaf from the rest of the plant. Ethene **stimulates** the cells in the abscission layer to **expand**, **breaking** the **cells walls** and causing the **leaf** to **fall off**.

Auxins are antagonistic to ethene.

Plant Responses

Plant Hormones have Many Commercial Uses

The **fruit industry** uses different **plant hormones** to **control** how different fruits develop, e.g:

Ethene stimulates the ripening of fruit

Ethene stimulates enzymes that **break down cell walls**, **break down chlorophyll** and convert **starch** into **sugars**. This makes the fruit **soft**, **ripe** and **ready to eat**.

E.g. **bananas** are harvested and transported **before** they're **ripe** because they're **less likely** to be **damaged** this way. They're then **exposed** to **ethene** on arrival so they **all ripen** at the **same time** on the **shelves** and in people's **homes**.

Auxins and gibberellins make fruit develop

Auxins and gibberellins are **sprayed** onto **unpollinated flowers**, which makes the **fruit develop without fertilisation**.

E.g. **seedless grapes** can be produced using **auxins** and **gibberellins**.

Auxins can prevent or trigger fruit drop

Applying a **low concentration** of auxins in the **early stages** of fruit production **prevents** the **fruit** from **dropping** off the plant. But applying a **high concentration** of auxins at a **later stage** of fruit production **triggers** the fruit to **drop**.

E.g. **apples** can be made to **drop off** the tree at **exactly** the **right time**.

Auxins are also used **commercially** by **farmers** and **gardeners**, for example:

Auxins are used in **selective weedkillers (herbicides)** — auxins make **weeds** produce **long stems** instead of lots of **leaves**. This makes the weeds **grow too fast**, so they **can't** get enough **water** or **nutrients**, so they **die**.

Auxins are used as **rooting hormones** — auxins make a **cutting** (part of the plant, e.g. a stem cutting) **grow roots**. The **cutting** can then be **planted** and **grown** into a new plant. **Many cuttings** can be taken from **just one original plant** and **treated** with **rooting hormones**, so **lots** of the same plant can be grown **quickly** and **cheaply** from just one plant.

Practice Questions

Q1 Give two reasons why plants need to respond to stimuli.

Q2 What is a tropism?

Q3 What is a plant growth hormone?

Q4 Give one function of gibberellins in a plant.

Q5 Which hormone inhibits leaf loss in deciduous plants?

Exam Questions

Q1 Explain how the movement of auxins in a growing shoot enables the plant to grow towards the light. [3 marks]

Q2 A gardener notices that one of the plants in his garden is showing apical dominance.
a) Name the type of plant hormone that controls apical dominance. [1 mark]
b) Give two advantages of apical dominance. [2 marks]

Q3 A tomato grower wants all her tomatoes to ripen at the same time, just before she sells them at a market.
a) Name a plant hormone she could use to make the tomatoes ripen. [1 mark]
b) Explain how the plant hormone named in part a) makes tomatoes ripen. [1 mark]
c) Suggest a commercial advantage of being able to pick and transport tomatoes before they're ripe. [1 mark]

The weeping willow — yep, that plant definitely has hormones...

See, told you plant responses were important — I didn't say exciting, I said important. Just wait till the next time you're in a supermarket — I bet you can't get round the whole shop without commenting on why the bananas are ripe...

Animal Responses

I'm afraid you're not seeing things — there's a little bit more about nervous and hormonal communication in this unit.

Responding to their *Environment* Helps *Animals Survive*

1) **Animals increase** their **chances** of **survival** by **responding** to **changes** in their **external environment**, e.g. by **avoiding harmful environments** such as places that are too hot or too cold.
2) They also **respond** to **changes** in their **internal environment** to make sure that the **conditions** are always **optimal** for their **metabolism** (all the chemical reactions that go on inside them).
3) Any **change** in the internal or external **environment** is called a **stimulus**.

You might remember a lot of this from Unit 4, but you have to know some of it for Unit 5 as well.

The *Nervous* and *Hormonal Systems Coordinate Responses*

1) **Receptors detect stimuli** and **effectors** bring about a **response** to a **stimulus**. Effectors include **muscle cells** and cells found in **glands**, e.g. the **pancreas**.
2) Receptors **communicate** with effectors via the **nervous system** or the **hormonal system**, or sometimes using **both**.

The *Nervous System* is *Split* into *Two Main Systems*

The **central nervous system** (**CNS**) — made up of the **brain** and the **spinal cord**.

The **peripheral nervous system** — made up of the neurones that connect the CNS to the **rest** of the **body**. It also has two different systems:

- The **somatic nervous system** controls **conscious** activities, e.g. running and playing video games.
- The **autonomic nervous system** controls **unconscious** activities, e.g. digestion. It's got two divisions that have **opposite effects** on the body:
  - The **sympathetic** nervous system gets the body **ready for action**. It's the '**fight or flight**' system. Sympathetic neurones release the neurotransmitter **noradrenaline**.
  - The **parasympathetic** nervous system **calms** the body down. It's the '**rest and digest**' system. Parasympathetic neurones release the neurotransmitter **acetylcholine**.

The *Brain* is *Part* of the *Central Nervous System*

You need to know the **location** and **function** of these **four brain structures**:

1 Cerebrum — Allows You to See, Hear, Learn and Think

1) The **cerebrum** is the **largest** part of the brain.
2) It's divided into **two halves** called **cerebral hemispheres**.
3) The cerebrum has a thin **outer layer** called the **cerebral cortex**, which is highly **folded**.
4) The cerebrum is involved in **vision**, **hearing**, **learning** and **thinking**.

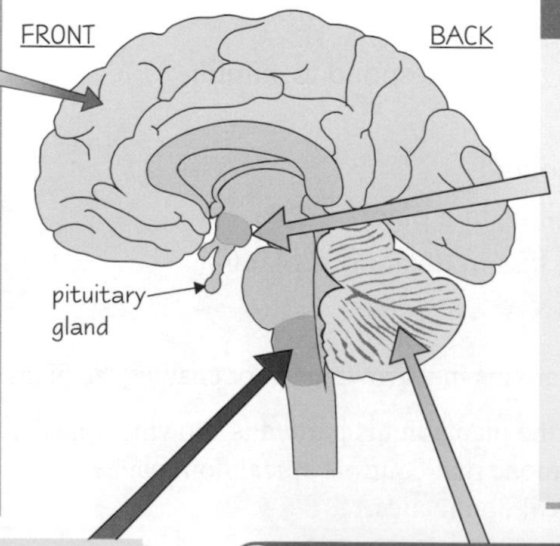

FRONT
BACK
pituitary gland

2 Hypothalamus — Controls Body Temperature

1) The hypothalamus is found just **beneath** the **middle part** of the brain.
2) The hypothalamus automatically **maintains body temperature** at the normal level (see p. 15).
3) The hypothalamus produces **hormones** that **control** the **pituitary gland** — a gland just below the hypothalamus.

4 Medulla Oblongata — Controls Breathing Rate and Heart Rate

1) The **medulla oblongata** is at the **base** of the **brain**, at the top of the spinal cord.
2) It automatically controls **breathing rate** and **heart rate**.

3 Cerebellum — Coordinates Muscles, Balance and Posture

1) The **cerebellum** is **underneath** the **cerebrum** and it also has a **folded cortex**.
2) It's important for **muscle coordination**, **posture** and **coordination of balance**.

Animal Responses

The Nervous and Hormonal Systems Coordinate the 'Fight or Flight' Response

1) When an organism is **threatened** (e.g. by a predator) it responds by **preparing the body for action** (e.g. for fighting or running away). This **response** is called the **'fight or flight'** response.

2) The **nervous system** and **hormonal system coordinate** the fight or flight response.

3) The **sympathetic** nervous system is **activated**, which also **triggers** the **release** of **adrenaline**. The sympathetic nervous system and adrenaline have the following effects:

Harold thought it was about time his sympathetic nervous system took over.

- **Heart rate** is **increased** — so **blood** is **pumped** around the body **faster**.
- The **muscles** around the **bronchioles relax** — so **breathing is deeper**.
- **Glycogen** is **converted** into **glucose** — so **more glucose** is **available** for **muscles** to **respire**.
- **Muscles** in the **arterioles** supplying the **skin** and **gut constrict**, and muscles in the **arterioles** supplying the **heart**, **lungs** and **skeletal muscles dilate** — so **blood** is **diverted** from the skin and gut **to** the **heart**, **lungs** and **skeletal muscles**.

Practice Questions

Q1 Why do organisms respond to changes in their environment?

Q2 Which part of the nervous system controls unconscious activities?

Q3 What does the sympathetic nervous system do?

Q4 Which part of the brain is involved in learning?

Q5 Which part of the brain controls body temperature?

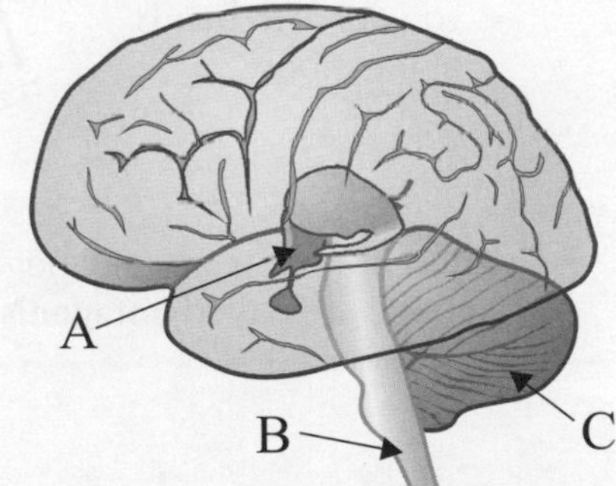

Exam Questions

Q1 The diagram on the right shows a cross-section of the brain from front to back.

a) Name structure A on the diagram of the brain. [1 mark]

b) Give two roles of structure B. [2 marks]

c) What effect might damage to structure C have on the body? [1 mark]

Q2 The nervous and hormonal systems coordinate the 'fight or flight' response.

a) What is the 'fight or flight' response? [1 mark]

b) Give an example of when it might occur. [1 mark]

c) Give one physiological effect of the response on the body. [1 mark]

The cere-mum part of the brain — coordinates dirty washing and clean clothes...

These pages aren't all bad — at least you've covered some of the information on them before, way back at the start of the book. The difference here is that you need to know more about how the nervous system is organised, the structure of that big old squelchy mess in your skull, and the fight or flight response — so make sure you test yourself on these new bits.

Muscle Contraction

Muscles are effectors — they contract so you can respond to your environment. You need to know how they contract, but first you need to know a bit more about them and how they're involved in movement...

The Central Nervous System (CNS) Coordinates Muscular Movement

1) The **CNS** (**brain** and **spinal cord**) receives **sensory information** and **decides** what kind of **response** is needed.
2) If the response needed is **movement**, the CNS sends signals along **neurones** to tell **skeletal muscles** to **contract**.
3) Skeletal muscle (also called striated, striped or voluntary muscle) is the type of muscle you use to **move**, e.g. the biceps and triceps move the lower arm.

Movement Involves Muscles, Tendons, Ligaments and Joints

1) **Skeletal muscles** are attached to **bones** by **tendons**.
2) **Ligaments** attach **bones** to **other bones**, to hold them together.
3) The **structure** of the **joints** between your bones determines what **kind** of **movement** is possible:
   - **Ball and socket joints** (e.g. the **shoulder**) allow movement in **all directions**.
   - **Gliding joints** (e.g. the **wrist**) allow a **wide range** of movement because small bones slide over each other.
   - **Hinge joints** (e.g. the **elbow**) allow movement in **one plane only**, like up and down.

Here's how your **muscles** work to **bend** your **arm** at the **elbow**:

- The bones of your **lower arm** are attached to a **biceps** muscle and a **triceps** muscle by **tendons**.
- The biceps and triceps **work together** to move your arm — as one **contracts**, the other **relaxes**:

When your **biceps contracts** your **triceps relaxes**. This pulls the bone so your **arm bends** at the elbow.

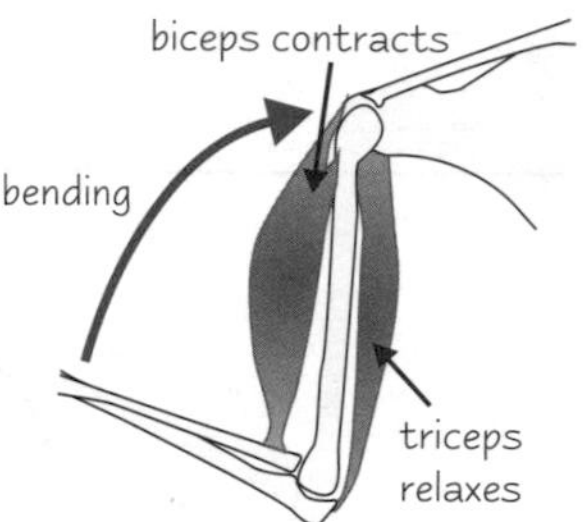

When your **triceps contracts** your **biceps relaxes**. This pulls the bone so your **arm straightens** at the **elbow**.

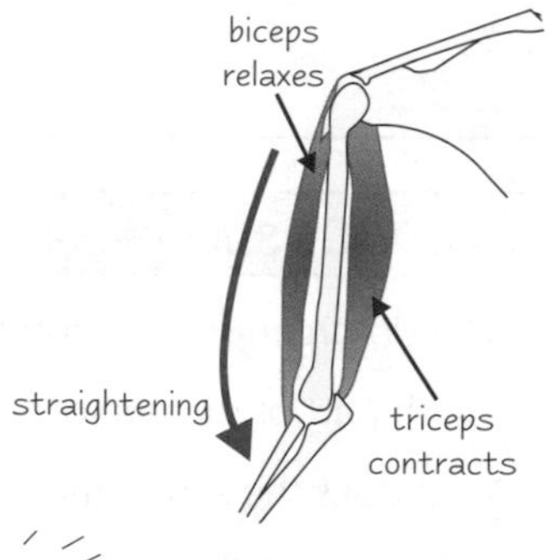

- Muscles that work together to move a bone are called **antagonistic pairs**.

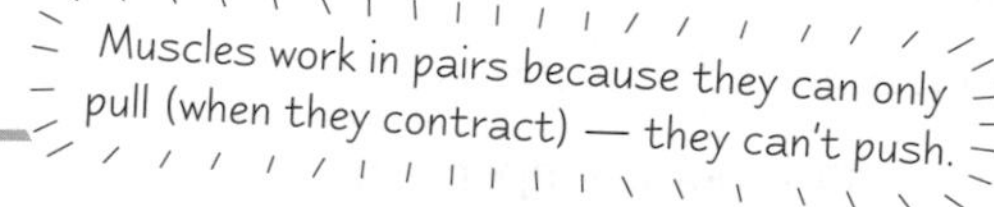

Skeletal Muscle is made up of Long Muscle Fibres

1) Skeletal muscle is made up of **large bundles** of **long cells**, called **muscle fibres**.
2) The cell membrane of muscle fibre cells is called the **sarcolemma**.
3) Bits of the sarcolemma **fold inwards** across the muscle fibre and stick into the **sarcoplasm** (a muscle cell's cytoplasm). These folds are called **transverse (T) tubules** and they help to **spread electrical impulses** throughout the sarcoplasm so they **reach** all parts of the **muscle fibre**.
4) A network of **internal membranes** called the **sarcoplasmic reticulum** runs through the sarcoplasm. The sarcoplasmic reticulum **stores** and **releases calcium ions** that are needed for muscle contraction (see p. 106).
5) Muscle fibres have lots of **mitochondria** to **provide** the **ATP** that's needed for **muscle contraction**.
6) They are **multinucleate** (contain many nuclei).
7) Muscle fibres have lots of **long**, **cylindrical organelles** called **myofibrils**. They're made up of proteins and are **highly specialised** for **contraction**.

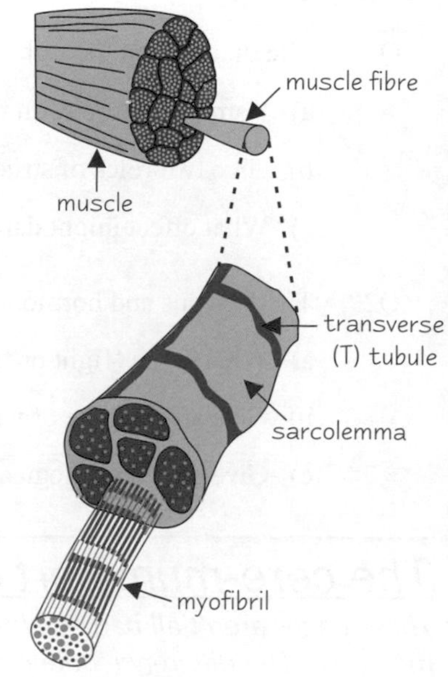

Muscle Contraction

Myofibrils Contain *Thick Myosin* Filaments and *Thin Actin* Filaments

1) Myofibrils contain bundles of **thick** and **thin myofilaments** that **move past each other** to make muscles **contract**.
   - **Thick myofilaments** are made of the protein **myosin**.
   - **Thin myofilaments** are made of the protein **actin**.
2) If you look at a **myofibril** under an **electron microscope**, you'll see a pattern of alternating **dark** and **light bands**:
   - **Dark** bands contain the **thick myosin filaments** and some overlapping thin actin filaments — these are called **A-bands**.
   - **Light** bands contain **thin actin filaments** only — these are called **I-bands**.
3) A myofibril is made up of many short units called **sarcomeres**.
4) The **ends** of each **sarcomere** are marked with a **Z-line**.
5) In the **middle** of each sarcomere is an **M-line**. The M-line is the **middle** of the **myosin** filaments.
6) **Around** the M-line is the **H-zone**. The H-zone **only** contains **myosin** filaments.

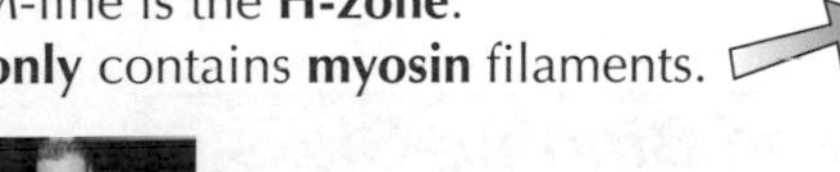

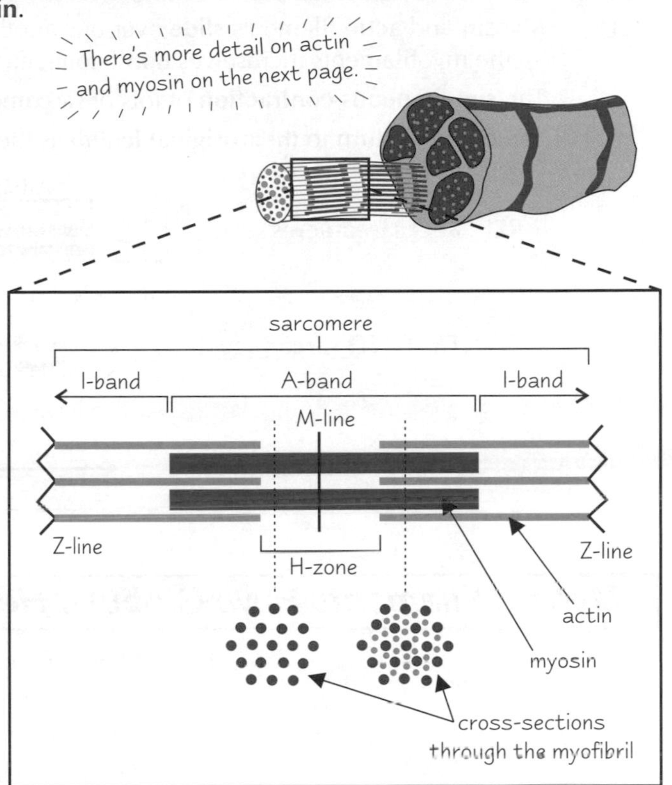

Derek was the proud winner of the biggest muscles AND the smallest pants.

Practice Questions

Q1 What is skeletal muscle?

Q2 When your biceps contracts, what happens to your triceps?

Q3 What are transverse (T) tubules?

Q4 Name the two proteins that make up myofibrils.

Q5 What are the light bands in a myofibril called?

Q6 Where is the M-line located in a sarcomere?

Exam Questions

Q1 Describe how myofilaments, muscle fibres, myofibrils and muscles are related to each other. [3 marks]

Q2 A muscle myofibril was examined under an electron microscope and a sketch was drawn (Figure 1).

a) What are the correct names for labels A, B and C? [3 marks]

b) The myofibril was then cut through the M-line (Figure 2). State which of the cross-section drawings you would expect to see and explain why. [3 marks]

Figure 1

A

B

C

Figure 2

1 2 3

Sarcomere — a French mother with a dry sense of humour...

Blimey, what a page. My head has A-bands, I-bands, what-bands and who-bands all swimming around in it. But I guess once you've learnt all these weird and wonderful names you'll never forget them — that's right, they'll take up vital brain space forever. But they'll also get you vital marks in your exam — providing you know what they all mean that is.

Muscle Contraction

Brace yourself — here comes the detail of muscle contraction...

Muscle Contraction is Explained by the Sliding Filament Theory

1) **Myosin** and **actin** filaments **slide** over one another to make the **sarcomeres contract** — the myofilaments themselves **don't** contract.
2) The **simultaneous contraction** of lots of **sarcomeres** means the **myofibrils** and **muscle fibres contract**.
3) Sarcomeres return to their **original length** as the muscle **relaxes**.

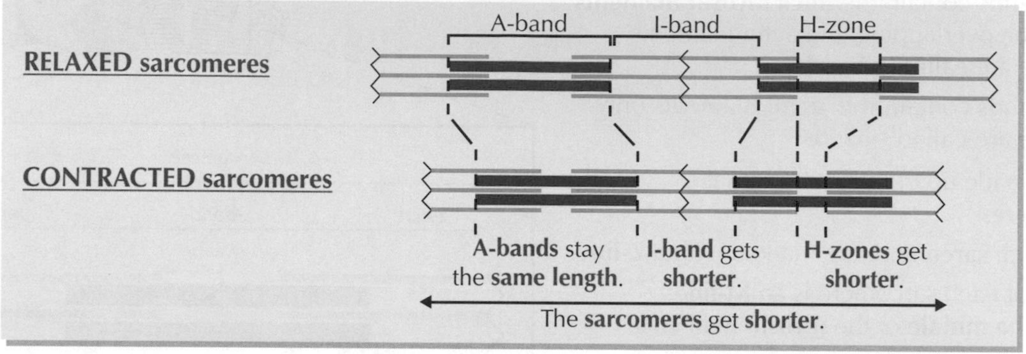

Myosin Filaments Have Globular Heads and Binding Sites

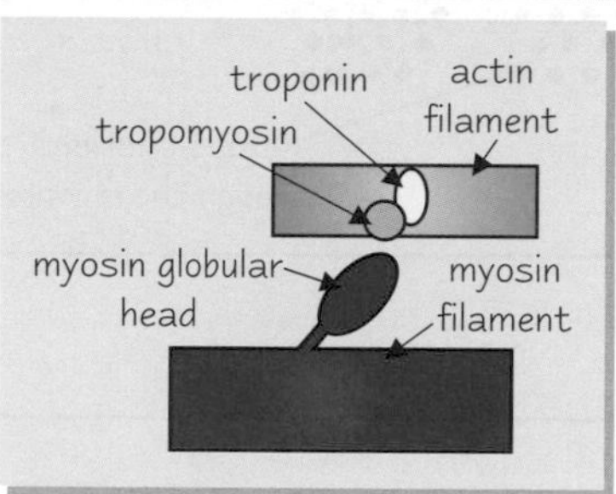

1) **Myosin filaments** have **globular heads** that are **hinged**, so they can move **back** and **forth**.
2) Each myosin head has a **binding site** for **actin** and a **binding site** for **ATP**.
3) **Actin filaments** have **binding sites** for **myosin heads**, called **actin-myosin** binding sites.
4) Two other **proteins** called **tropomyosin** and **troponin** are found between actin filaments. These proteins are **attached** to **each other** and they **help** myofilaments **move** past each other.

Binding Sites in Resting Muscles are Blocked by Tropomyosin

1) In a **resting** (unstimulated) muscle the **actin-myosin binding site** is **blocked** by **tropomyosin**, which is held in place by **troponin**.
2) So **myofilaments can't slide** past each other because the **myosin heads can't bind** to the actin-myosin binding site on the actin filaments.

Muscle Contraction is Triggered by an Action Potential

1 The Action Potential Triggers an Influx of Calcium Ions

1) When an action potential from a motor neurone **stimulates** a muscle cell, it **depolarises** the **sarcolemma**. Depolarisation **spreads** down the **T-tubules** to the **sarcoplasmic reticulum** (see p. 104).
2) This causes the **sarcoplasmic reticulum** to **release** stored **calcium ions** (Ca^{2+}) into the **sarcoplasm**.

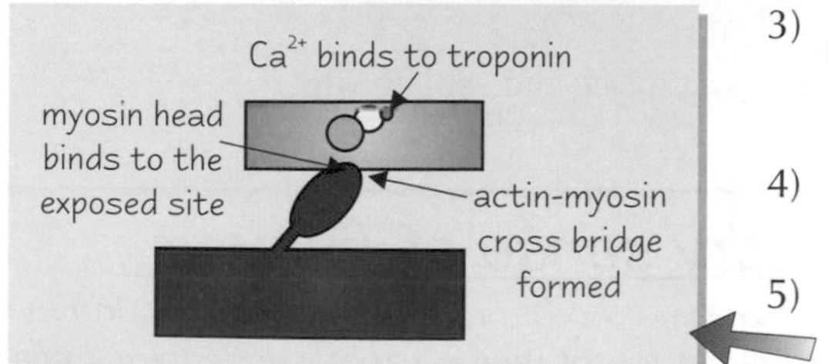

3) Calcium ions **bind** to **troponin**, causing it to **change shape**. This **pulls** the attached **tropomyosin out** of the **actin-myosin binding site** on the actin filament.
4) This **exposes** the **binding site**, which allows the **myosin head** to **bind**.
5) The bond formed when a **myosin head** binds to an **actin filament** is called an **actin-myosin cross bridge**.

Muscle Contraction

2 ATP Provides the Energy Needed to Move the Myosin Head...

1) **Calcium** ions also **activate** the enzyme **ATPase**, which **breaks down ATP** (into ADP + P_i) to **provide** the **energy** needed for muscle contraction.
2) The **energy** released from ATP **moves** the **myosin head**, which **pulls** the **actin filament** along in a kind of **rowing action**.

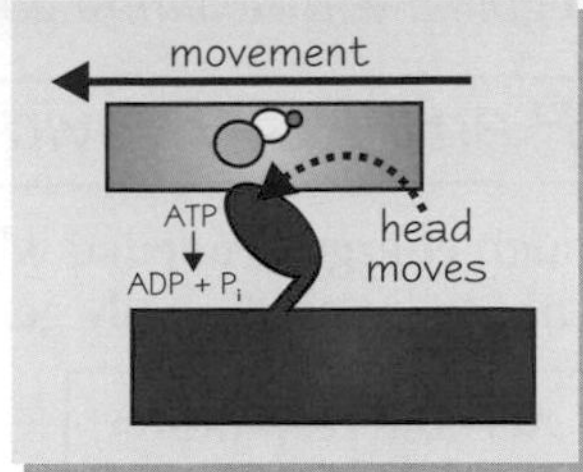

3 ...and to Break the Cross Bridge

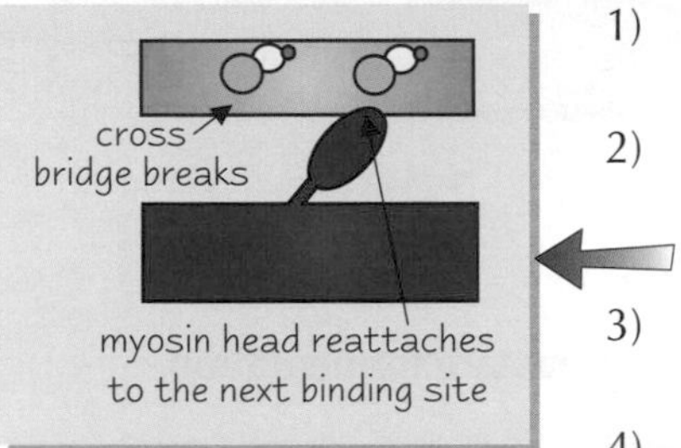

1) **ATP** also provides the **energy** to **break** the **actin-myosin cross bridge**, so the **myosin head detaches** from the actin filament **after** it's moved.
2) The **myosin head** then **reattaches** to a **different binding site** further along the actin filament. A **new actin-myosin cross bridge** is formed and the **cycle** is **repeated** (attach, move, detach, reattach to new binding site...).
3) **Many** cross bridges **form** and **break** very **rapidly**, pulling the actin filament along — which **shortens** the **sarcomere**, causing the **muscle** to **contract**.
4) The cycle will **continue** as long as **calcium ions** are **present** and **bound** to **troponin**.

When Excitation Stops, Calcium Ions Leave Troponin Molecules

1) When the muscle **stops** being **stimulated**, **calcium ions leave** their **binding sites** on the **troponin** molecules and are moved by **active transport** back into the **sarcoplasmic reticulum** (this needs **ATP** too).
2) The **troponin** molecules return to their **original shape**, pulling the attached **tropomyosin** molecules with them. This means the **tropomyosin** molecules **block** the actin-myosin **binding sites** again.
3) Muscles **aren't contracted** because **no myosin heads** are **attached** to **actin** filaments (so there are no actin-myosin cross bridges).
4) The **actin** filaments **slide back** to their **relaxed** position, which **lengthens** the **sarcomere**.

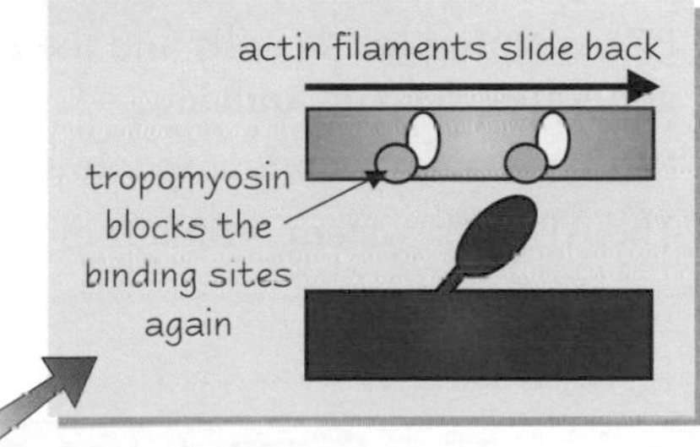

Practice Questions

Q1 What happens to sarcomeres as a muscle relaxes?

Q2 Which molecule blocks the actin-myosin binding site in resting muscles?

Q3 What's the name of the bond that's formed when a myosin head binds to an actin filament?

Exam Questions

Q1 Describe how the lengths of the different bands in a myofibril change during muscle contraction. [2 marks]

Q2 Rigor mortis is the stiffening of muscles in the body after death. It happens when ATP reserves are exhausted. Explain why a lack of ATP leads to muscles being unable to relax. [3 marks]

Q3 Bepridil is a drug that blocks calcium ion channels. Describe and explain the effect this drug will have on muscle contraction. [3 marks]

What does muscle contraction cost? 80p...

Sorry, that's my favourite sciencey joke so I had to fit it in somewhere — a small distraction before you revisit this page. It's tough stuff but you know the best way to learn it. That's right, shut the book and scribble down what you can remember — if you can't remember much, read it again till you can (and if you can remember loads read it again anyway, just to be sure).

Muscle Contraction

Keep going, you've almost got muscles done and dusted — just a few more bits and pieces to learn about them.

ATP and PCr Provide the Energy for Muscle Contraction

So much **energy** is **needed** when muscles contract that **ATP** gets **used up very quickly**.
ATP has to be **continually generated** so exercise can continue — this happens in **three main ways**:

① Aerobic respiration

- **Most ATP** is generated via **oxidative phosphorylation** in the cell's **mitochondria**.
- **Aerobic** respiration only works when there's **oxygen** so it's good for **long periods** of **low-intensity exercise**, e.g. walking or jogging.

See pages 38-45 for more on aerobic and anaerobic respiration.

② Anaerobic respiration

- ATP is made **rapidly** by **glycolysis**.
- The **end product** of glycolysis is **pyruvate**, which is converted to **lactate** by **lactate fermentation**.
- Lactate can **quickly build up** in the muscles and cause **muscle fatigue**.
- Anaerobic respiration is good for **short periods** of **hard exercise**, e.g. a **400 m sprint**.

③ ATP-Phosphocreatine (PCr) System

- **ATP** is made by **phosphorylating ADP** — adding a phosphate group taken from **phosphocreatine** (**PCr**).
- **PCr is stored** inside cells and the ATP-PCr system **generates ATP** very **quickly**.
- **PCr runs out** after a few seconds so it's used during **short bursts** of **vigorous exercise**, e.g. a **tennis serve**.
- The ATP-PCr system is **anaerobic** (it doesn't need oxygen) and it's **alactic** (it doesn't form any lactate).

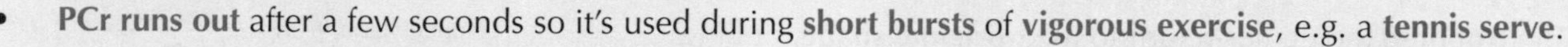

ADP + PCr → ATP + Cr (creatine)

Many activities use a combination of these systems.

There are Three Types of Muscle

❶ Voluntary muscle (skeletal muscle)

1) **Voluntary** muscle contraction is controlled **consciously** (you have to voluntarily decide to contract it).
2) It's made up of **many muscle fibres** that have **many nuclei**.
3) The muscle fibres can be **many centimetres long**.
4) You can see regular **cross-striations** (a **striped pattern**) under a **microscope**.
5) Some muscle fibres **contract very quickly** — they're used for **speed** and **strength** but **fatigue** (get tired) **quickly**.
6) Some muscle fibres **contract slowly** and **fatigue slowly** — they're used for **endurance** and **posture**.

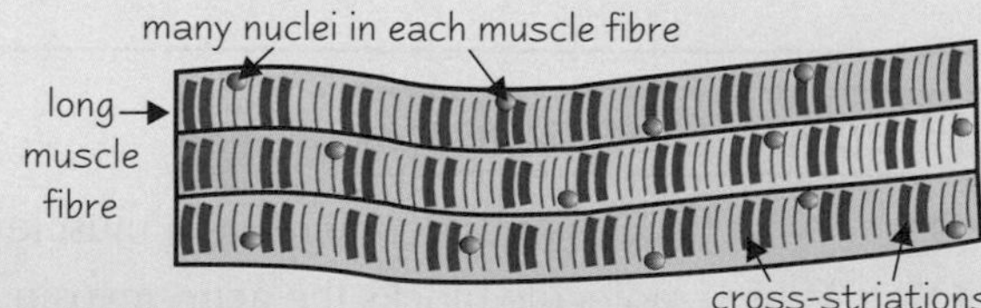

❷ Involuntary muscle (also called smooth muscle)

1) **Involuntary** muscle contraction is controlled **unconsciously** (it'll contract automatically without you deciding to).
2) It's also called **smooth muscle** because it **doesn't** have the **striped appearance** of voluntary muscle.
3) It's found in the **walls** of your **hollow internal organs**, e.g. the **gut**, the **blood vessels**. Your **gut smooth muscles contract** to **move food along** (**peristalsis**) and your **blood vessel smooth muscles contract** to **reduce** the **flow** of **blood**.
4) Each muscle fibre has **one nucleus**.
5) The muscle fibres are **spindle-shaped** with **pointed ends**, and they're only about **0.2 mm long**.
6) The muscle fibres **contract slowly** and **don't fatigue**.

nucleus
spindle-shaped muscle fibre

Muscle Contraction

3 Cardiac muscle (heart muscle)

1) **Cardiac** muscle **contracts** on its **own** — it's **myogenic** (but the **rate** of contraction is controlled involuntarily by the **autonomic nervous system**).
2) It's found in the **walls** of your **heart**.
3) It's made of muscle fibres **connected** by **intercalated discs**, which have **low electrical resistance** so nerve impulses pass **easily** between cells.
4) The muscle fibres are **branched** to allow **nerve impulses** to **spread quickly** through the whole muscle.
5) Each muscle fibre has **one nucleus**.
6) The muscle fibres are shaped like **cylinders** and they're about **0.2 mm long**.
7) You can see **some cross-striations** but the striped pattern **isn't** as **strong** as it is in voluntary muscle.
8) The muscle fibres **contract rhythmically** and **don't fatigue**.

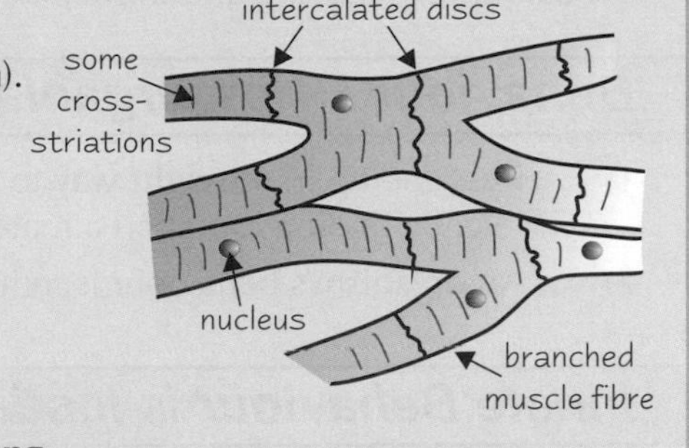

Neuromuscular Junctions are Synapses Between Neurones and Muscles

1) A **neuromuscular junction** is a **synapse** between a **motor neurone** and a **muscle cell**.
2) Neuromuscular junctions use the neurotransmitter **acetylcholine** (**ACh**), which binds to receptors called **nicotinic cholinergic receptors**.
3) Neuromuscular junctions **work** in the **same way** as **synapses between neurones** — they **release neurotransmitter**, which triggers **depolarisation** in the **postsynaptic cell** (see pages 10-11).

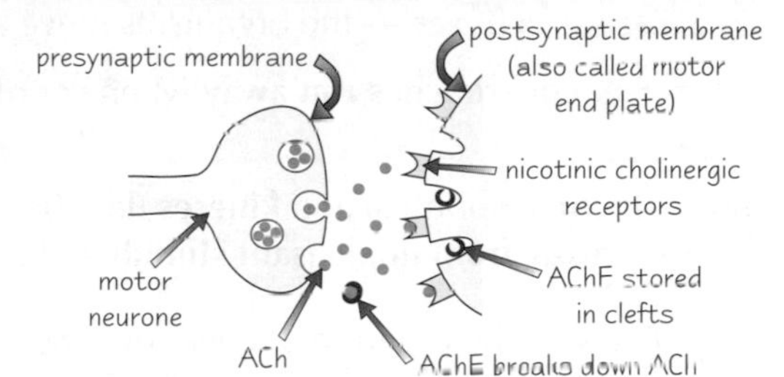

4) There are a few **differences** between neuromuscular junctions and synapses where two neurones meet:

	NEUROMUSCULAR JUNCTIONS	SYNAPSES (between neurones)
Neurotransmitter	Acetylcholine	Various
Postsynaptic receptors	Nicotinic cholinergic receptors	Various
Number of postsynaptic receptors	Lots	Fewer
Postsynaptic cell	Muscle cell	Neurone
Postsynaptic membrane	Has clefts containing AChE	Smooth
Effect of neurotransmitter binding to postsynaptic receptors	Muscle cell always contracts	Action potential may or may not fire in the next neurone
Removal of neurotransmitter	Broken down by AChE	Various ways (it depends on the neurotransmitter)

Practice Questions

Q1 What type of respiration provides energy for long periods of low-intensity exercise?

Q2 Which type of muscle has many nuclei in each muscle fibre?

Exam Questions

Q1 Compare the structure and function of involuntary muscle and cardiac muscle. [5 marks]

Q2 Describe three differences between neuromuscular junctions and synapses. [3 marks]

Smooth muscle — it has a way with the ladies...

There's a lot of information on these two pages (not unlike the past few) but you just have to sit down and learn it — there's no other way I'm afraid. But at least that's it for muscle contraction — now you just need to get through behaviour.

Behaviour

Behaviour is an organism's response to changes in its external environment. I thought there were only two types of behaviour — behaving and misbehaving. Turns out there are a few more...

Behaviour Helps **Organisms** to **Survive** and **Reproduce**

1) Responding in the **right way** to an environmental change helps organisms **survive** and **reproduce** (e.g. by finding food and a mate).
2) An organism's behaviour is influenced by both its **genes** and its **environment**.

Innate Behaviour *is* ***Instinctive*** *and* ***Inherited***

1) **Innate** behaviour is **behaviour** that organisms do **instinctively**.
2) It's **genetically determined** — it's **inherited** from parents and it's **not** influenced by the **environment**.
3) It's also **stereotyped** — it's always carried out in the **same way** and by **all** the **individuals** in a species.
4) The **advantage** of innate behaviour is that organisms **respond** in the **right way** to the stimulus **straight away** because **no learning** is needed, e.g. newborn babies instinctively suckle from their mothers.
5) You need to know **three examples** of **innate behaviours**:

① **Escape reflexes** — the organisms **move away** from **potential danger**.

E.g. **cockroaches run away** when your foot's about to squash them.

Dave's escape reflex was about to kick in.

Taxes (**tactic** responses) and **kineses** (**kinetic** responses) allow simple organisms to move **away from unpleasant stimuli** and into **more favourable** environments:

② **Taxes** — the organisms move **towards** or **away from** a **directional stimulus**.

E.g. **woodlice** move **away from** a **light source**. This helps keep them **concealed** under stones where they're **safe** from predators, so it helps them **survive**.

③ **Kineses** — the organisms' **movement** response is affected by a **non-directional stimulus**, e.g. **intensity**.

E.g. **woodlice** show a **kinetic** response to **low humidity**. This helps them **move** from **drier air** to more **humid air**, and then **stay put**. This **reduces** their **water loss** so **improves** their **survival** chances.

Learned Behaviour *is* ***Behaviour*** *that's* ***Modified*** *as a Result of* ***Experience***

Learned behaviour is **influenced** by the **environment**. It allows animals to **respond** to **changing conditions**, e.g. they **learn** to **avoid harmful food**. Here are some **examples** that you need to **learn**:

1 Habituation

- **Habituation** is a **reduced response** to an **unimportant stimulus** after **repeated** exposure **over time**.
- An **unimportant stimulus** is a change that **isn't threatening or rewarding**. An animal quickly learns to **ignore** it so it **doesn't waste time** and **energy** responding to unimportant things, e.g. you **learn** to **sleep through traffic noise** at night.
- Animals **remain alert** to **unfamiliar stimuli** though, e.g. you instantly **wake up** if you hear an **unfamiliar noise**.

2 Classical Conditioning

- **Classical conditioning** is **learning** to **respond naturally** to a **stimulus** that **doesn't normally** cause that response.
- A **natural stimulus** (called the **unconditioned** stimulus) can cause a **natural response** (called the unconditioned response). E.g. in dogs, **food** (an **unconditioned** stimulus) causes **salivation** (an unconditioned response).
- If another stimulus **coincides** with an **unconditioned** stimulus **enough times**, eventually this other stimulus will cause the **same response**. E.g. if a **bell** is rung **immediately before** dogs are given **food**, after a **time** the dogs will learn to **salivate** in **response** to the **bell only**.

The behaviour is automatic.

Behaviour

3 Operant Conditioning

- **Operant conditioning** is **learning** to **associate** a particular **response** with a **reward** or a **punishment**.
- When put in the **same situation lots of times**, an animal will work out **which response** gets a **reward** (e.g. pressing the right lever gets food) or a **punishment** (e.g. pressing the left lever gives a shock).
- The response must be **rewarded** (or punished) **straight away** — this **reinforces** the animal's behaviour so it's **more likely** to respond in the **same way** to get the **reward again** (or less likely to do it to be punished again).

E.g. a **rat** was put in a cage with a **choice** of **levers**. Pressing one of the levers **rewarded** the rat with food **straight away**. The rat was **repeatedly** put in the **same cage**, so learned which **lever** to **press** to get the **reward**.

- Lots of mistakes are made at first, but animals **quickly learn** to make **fewer mistakes** by using **trial and error**.

4 Latent Learning

- **Latent** learning is **hidden** learning — an animal **doesn't immediately show** it's learned something.
- It involves **learning** through **repeatedly** doing the **same task**.
- The animal only **shows** it's learned something when it's given a **reward** or a **punishment**.

E.g. **three groups** of **rats** were **repeatedly** put in the **same maze**:

1) The first group of rats were **reinforced** with a **reward** each time they reached the **end** of the **maze** — they **quickly learned** their way around the maze.
2) The second group of rats were **not reinforced** (they didn't receive a reward) — they continued to plod about the maze and **took ages** to reach the end.
3) The third group of rats were **only rewarded** from the **11th time** they did the maze — after this they **very quickly reached the end**, with **hardly any errors**. The rats had been **learning** the maze all along **without reinforcement**, but they **didn't show** their learning **until** there **was** a **reward**.

5 Insight Learning

- **Insight** is learning to **solve** a **problem** by **working out** a **solution** using **previous experience**.
- Solving problems by **insight** is **quicker** than by trial and error because actions are **planned** and **worked out**.

E.g. **chimpanzees** were put in a play area with **sticks**, **clubs** and **boxes**. Bunches of **bananas** were hung just **out of reach**. The chimps used their **previous experience** of playing with the objects to **work out** a **solution** — they **piled up** the boxes to reach the bananas, and used the sticks and clubs to **knock them down**.

Practice Questions

Q1 What is an escape reflex?

Q2 What's the difference between taxes and kineses?

Q3 What is habituation?

Exam Questions

Q1 When a postman puts a letter through Number 10, a dog barks loudly causing the postman's heart rate to increase. This happens for the next few days until the postman's heart rate increases as he approaches Number 10, even if the dog is not there. State what type of learning has occurred and give a reason for your answer. [2 marks]

Q2 Give an example of how operant conditioning could be used in dog training. [2 marks]

My hair's so shiny — it's classically conditioned...

Behaviour is a bit of a weird topic, really. It seems you can only find anything out by doing lots of strange experiments. Might be fun to work in this area of research though — it's certainly not your average day at the office. "How was your day, dear?" "Great thanks — the rats sped around the maze to get their chocolate." And who said science was boring...

Behaviour

All this studying of animal behaviour helps us to understand human behaviour. But animals are a bit different from us. Not many dogs study for exams, for a start. Mind you, they don't need A2 levels — they've already got a pe-degree...

Imprinting is a Combination of Innate and Learned Behaviour

1) **Imprinting** is a combination of a **learned** behaviour and an **innate** behaviour — e.g. an animal **learns to recognise its parents**, and **instinctively follows them**.
2) Imprinting occurs in several species, mainly **birds**, which are **able to move** very **soon** after they're **born**. A newly-born animal has an **innate instinct** to **follow** the **first moving object** it sees — usually this would be its **mother** or **father**, who would **provide warmth**, **shelter** and **food** (helping it to survive).
3) But the animal has **no innate instinct** of what its parents **look like** — they have to **learn** this.
4) Imprinting **only happens** during a certain period of time **soon after** the animal is **born**. This period of time is called the **critical period**.

> E.g. ducklings usually imprint on their parent ducks. But if ducklings are **reared from birth** (during the **critical period**) by a **human**, then the human is the **first moving object** the ducklings **see** — so the ducklings **imprint** on the human (they follow them).

"Who's your daddy..."

5) Once learned, imprinting is **fixed** and **irreversible**. Animals use imprinting later in life to **identify mates** from the **same species**.

The Dopamine Receptor D_4 is Linked to Human Behaviour

1) An animal's **behaviour** depends on the **structure** and **function** of its **brain** (e.g. neurotransmitters, synapses, receptors, etc.).
2) Even fairly **small differences** in the brain can produce **big differences** in **behaviour**.
3) Much of our **understanding** of **human behaviour** comes from **studying** people with **abnormal behaviour**, to see how their **brains** are **different** from the brains of people who behave 'normally'.
4) Any **differences** in the brain give scientists **clues** to understanding how normal behaviour is **controlled**. For example:

- The **D_4 receptor** is a receptor in the brain for a neurotransmitter called **dopamine**.
- Having **too many D_4 receptors** in the brain has been **linked** to **abnormal behaviour**, e.g. the abnormal behaviour seen in **schizophrenia** — a disorder that affects **thinking**, **perception**, **memory** and **emotions**.
- The **evidence** for this **link** includes:
  1. If **drugs** that **stimulate** dopamine receptors are given to **healthy people**, it **causes** the **abnormal behaviour** seen in **schizophrenia**.
  2. **Drugs** that **block** D_4 receptors **reduce symptoms** in people with schizophrenia.
  3. People with **schizophrenia** have a **higher density** of **D_4 receptors** in their brain.
  4. One of the drugs that's used to **treat** schizophrenia **binds** to D_4 receptors **better** than it binds to other dopamine receptors.
- The **link** between the **D_4 receptor** and **abnormal behaviour** helps us to understand the **role** that the **D_4 receptor** plays in **normal behaviour**, e.g. it's involved in **thinking**, **perception**, **memory** and **emotions**.

See p. 10 if you can't remember about neurotransmitters and their receptors.

The D4 receptor protein is made by the DRD4 gene.

Behaviour

Social Behaviour in Primates has Many Advantages

Many animals **live together** in large **groups**. Behaviour that involves members of the group **interacting** with each other is called **social behaviour**. The **primates** (e.g. baboons, apes, humans) have more developed **social behaviour** than other animals. Social behaviour has many **advantages**. Here are some **examples** of **social behaviour** in **baboons** and the **advantages** of the behaviours:

1) **Baboons** live in groups, with about 50 baboons in each group.

   A large group like this is **more efficient** at **hunting** for **food** — together the baboons can **search** a **large area** and **communicate** back to the group where there's a good source of **food**.

2) Within each group there's a **clear-cut hierarchy** of **adult males**.

   This helps to **prevent fighting** (which **wastes energy**) because the males already **know** their **rank order** in the group.

The kids knew they'd have to move up the hierarchy before they could enjoy their go on the slide.

3) As each group **moves through** its own **territory** hunting for food, baboons **cooperate** with each other — **infant baboons** stay with their **mother** in the **middle** of the group and the **adult males** stay on the **outside** of the group.

   Infants and the **females** are **protected** if they're on the **inside** of the group. The young baboons need to be kept **safe** and there needs to be **enough female baboons** for the males to **mate** with, to make sure that **reproduction** is **successful** and the group continues.

4) Members of the group **groom** each other (they **pick out small insects** and **dirt** from each other's **fur**).

   Grooming is **hygienic** and helps to **reinforce** the **social bonds** within the group.

Practice Questions

Q1 Give an example of a species that shows imprinting behaviour.

Q2 Can imprinting be reversed?

Q3 Name one behaviour that's linked to the D_4 receptor.

Q4 Give one example of social behaviour in primates.

Exam Questions

Q1 Goslings usually imprint on their parent geese. However, it's possible for a gosling to imprint on a human.

a) Explain what is meant by imprinting. [1 mark]

b) Explain how a gosling can imprint on a human. [2 marks]

Q2 Gorillas eat leaves, fruits and bark. They usually live in groups of 8-12 individuals. They exhibit many social behaviours that have many advantages.

a) Describe what is meant by the term 'social behaviour'. [1 mark]

b) Suggest a possible advantage to gorillas of:
   i) working together to look for food. [1 mark]
   ii) grooming each other. [1 mark]

Dopamine — wasn't he one of the seven dwarfs...

More crazy behaviour stuff. Apparently, a duckling will imprint to almost anything that moves about and makes a noise — try it with a football next time you're near a pond. No, don't really. It'll only go and disturb every match you play. So forget the practical, concentrate on the theory — learn these pages and then you've finished the book. Give me a whoop whoop...

How to Interpret Experiment and Study Data

Science is all about getting good evidence to test your theories... so scientists need to be able to spot a badly designed experiment or study a mile off, and be able to interpret the results of an experiment or study properly. Being the cheeky little monkeys they are, your exam board will want to make sure you can do it too. Here's a quick reference section to show you how to go about interpreting data-style questions.

Here Are Some Things You Might be Asked to do...

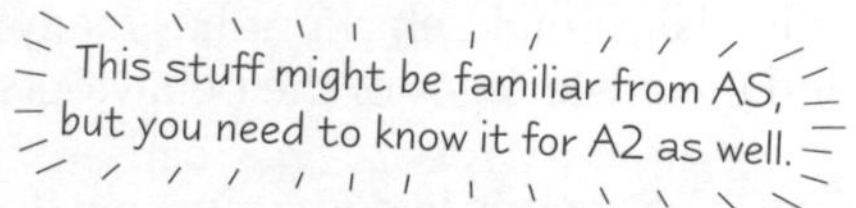

Here are three examples of the kind of data you could expect to get:

Study A

An agricultural scientist investigated the effect of three different pesticides on the number of pests in wheat fields. The number of pests was estimated in each of three fields, using ground traps, before and 1 month after application of one of the pesticides. The number of pests was also estimated in a control field where no pesticide had been applied. The table shows the results.

Pesticide	Number of pests	
	Before application	1 month after application
1	89	98
2	53	11
3	172	94
Control	70	77

Study B

Study B investigated the link between the number of bees in an area and the temperature of the area. The number of bees was estimated at ten 1-acre sites. The temperature was also recorded at each site. The results are shown in the scattergram below.

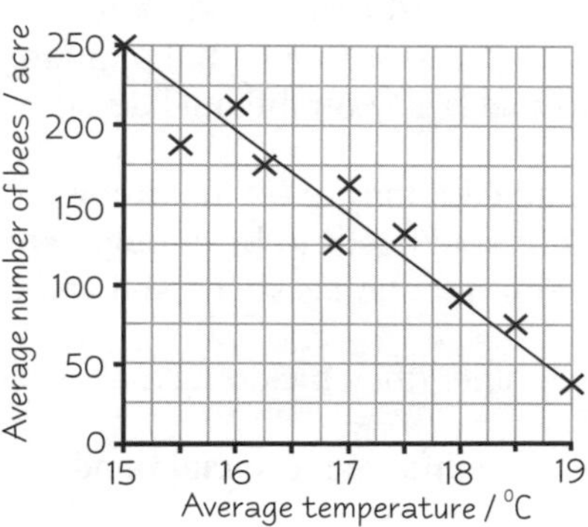

Experiment C

An experiment was conducted to investigate the effect of temperature on the rate of photosynthesis. The rate of photosynthesis in Canadian pondweed was measured at four different temperatures by measuring the volume of oxygen produced. All other variables were kept constant. The results are shown in the graph below.

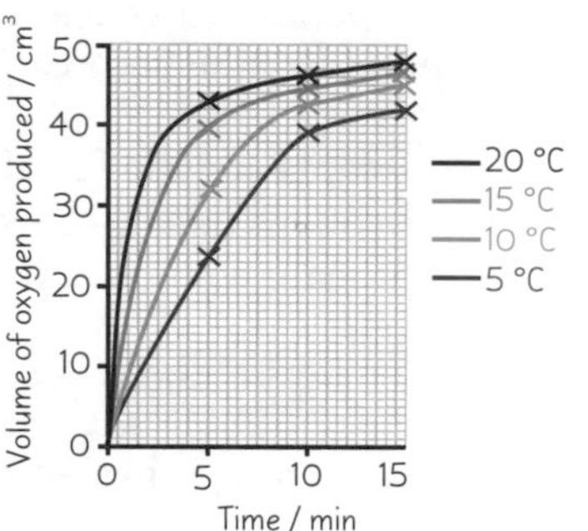

1) Describe and Manipulate the Data

You need to be able to **describe** any data you're given. The level of **detail** in your answer should be appropriate for the **number of marks** given. Loads of marks = more detail, few marks = less detail. You could also be asked to **manipulate** the data you're given (i.e. do some **calculations** on it). For the examples above:

Example — Study A

1) You could be asked to **calculate** the **percentage change** (**increase** or **decrease**) in the number of pests for each of the pesticides and the control. E.g. for pesticide 1: (98 – 89) ÷ 89 = 0.10 = **10% increase**.
2) You can then use these values to **describe** what the **data** shows — the **percentage increase** in pests in the field treated with **pesticide 1 was the same as for the control** (10% increase) (1 mark). **Pesticide 3 reduced** pest numbers by **45%**, but **pesticide 2** reduced the pest numbers the **most** (79% decrease) (1 mark).

Example — Study B

The data shows a **negative correlation** between the average number of bees and the temperature (1 mark).

Correlation describes the **relationship** between two variables — e.g. the one that's been changed and the one that's been measured. Data can show **three** types of correlation:

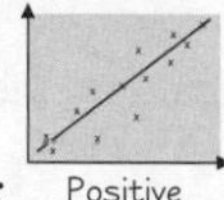
Positive

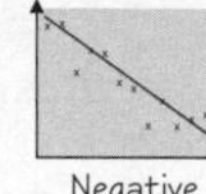
Negative

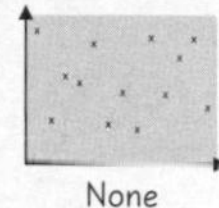
None

1) **Positive** — as one variable **increases** the other **increases**.
2) **Negative** — as one variable **increases** the other **decreases**.
3) **None** — there is **no relationship** between the two variables.

To tell if some data in a table **is correlated** — draw a **scatter diagram** of one variable against the other and **draw a line of best fit**.

Example — Experiment C

You could be asked to calculate the initial rate of photosynthesis at each temperature: The **gradient = the rate of photosynthesis**:

$$\text{Gradient} = \frac{\text{Change in Y}}{\text{Change in X}}$$

How to Interpret Experiment and Study Data

2) Draw or Check a Conclusion

1) Ideally, only **two** quantities would ever change in any experiment or study — everything else would be **constant**.
2) If you can keep everything else constant and the results show a correlation then you **can** conclude that the change in one variable **does cause** the change in the other.
3) But usually all the variables **can't** be controlled, so other **factors** (that you **couldn't** keep constant) could be having an **effect**.
4) Because of this, scientists have to be very careful when **drawing conclusions**. Most results show a **link** (**correlation**) between the variables, but that **doesn't prove that a change in one causes the change in the other.**

Example — Experiment C

All other variables were **kept constant**. E.g. light intensity and CO_2 concentration **stayed the same** each time, so these **couldn't** have influenced the rate of reaction. So you **can say** that an increase in temperature up to 20 °C **causes** an increase in the rate of photosynthesis.

Example — Study B

There's a **negative correlation** between the average number of bees and temperature. But you **can't** conclude that the increase in temperature **causes** the decrease in bees. **Other factors** may have been involved, e.g. there may be **less food** in some areas, there may be **more bee predators** in some areas, or **something else** you hadn't thought of could have caused the pattern...

Example — Experiment C

A science magazine **concluded** from this data that the optimum temperature for photosynthesis is **20 °C**. The data **doesn't** support this. The rate **could** be greatest at 22 °C, or 18 °C, but you can't tell from the data because it doesn't go **higher** than 20 °C and **increases** of **5 °C** at a time were used. The rates of photosynthesis at in-between temperatures **weren't** measured.

5) The **data** should always **support** the conclusion. This may sound obvious but it's easy to **jump** to conclusions. Conclusions have to be **precise** — not make sweeping generalisations.

3) Explain the Evidence

You could also be asked to **explain** the **evidence** (the data and results) — basically use your **knowledge** of the subject to explain **why** those results were obtained.

Example — Experiment C

Temperature increases the rate of photosynthesis because it **increases** the **activity** of **enzymes** involved in photosynthesis, so reactions are catalysed more quickly.

4) Comment on the Reliability of the Results

Reliable means the results can be **consistently reproduced** when an experiment or study is repeated. And if the results are reproducible they're more likely to be **true**. If the data isn't reliable for whatever reason you **can't draw** a valid **conclusion**. Here are some of the things that affect the reliability of data:

1) **Size of the data set** — For experiments, the **more repeats** you do, the **more reliable** the data. If you get the **same result** twice, it could be the correct answer. But if you get the same result **20 times**, it's much more reliable. The general rule for **studies** is the larger the **sample size**, the more **reliable** the **data** is.

   E.g. Study B is quite **small** — they only studied ten 1-acre sites. The **trend** shown by the data may not appear if you studied **50 or 100 sites**, or studied them for a longer period of time.

2) **The range of values in a data set** — The **closer** all the values are to the **mean**, the **more reliable** the data set.

   E.g. Study A is **repeated three more times** for pesticides 2 and 3. The percentage decrease each time is: 79%, 85%, 98% and 65% for **pesticide 2** (**mean = 82%**) and 45%, 45%, 54% and 43% for **pesticide 3** (**mean = 47%**). The data values are **closer to the mean** for **pesticide 3** than pesticide 2, so that data set is **more reliable**. The **spread** of **values about the mean** can be shown by calculating the **standard deviation** (SD).

The **smaller the SD** the **closer** the values to the **mean** and the **more reliable the data**. SDs can be shown on a graph using **error bars**. The ends of the bars show one SD **above** and one SD **below** the **mean**.

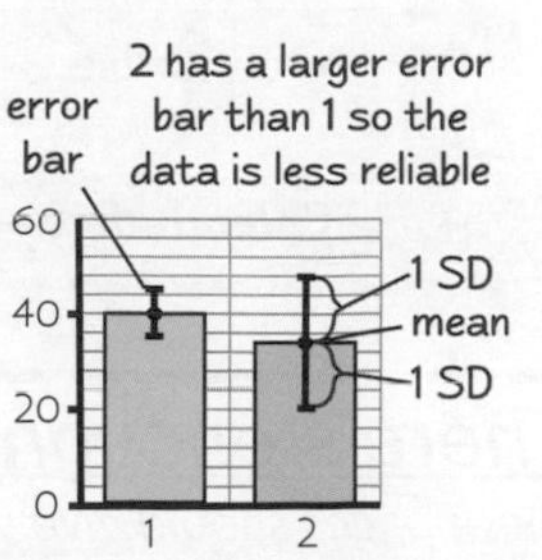

How to Interpret Experiment and Study Data

3) **Variables** — The **more variables** you **control**, the **more reliable** your data is. In an experiment you would control all the variables. In a study you try to control **as many as possible**.

E.g. ideally, all the sites in Study B would have a similar **type** of land, similar **weather**, have the same **plants** growing, etc. Then you could be more sure that the one factor being **investigated** (temperature) is having an **effect** on the thing being **measured** (number of bees).

The hat, trousers, shirt and tie variables had been well controlled in this study.

4) **Data collection** — think about all the **problems** with the **method** and see if **bias** has slipped in.

E.g. in Study A, the traps were placed on the **ground**, so pests like moths or aphids weren't included. This could have affected the results.

5) **Controls** — without controls, it's very difficult to **draw valid conclusions**. **Negative controls** are used to make sure that nothing you're doing in the experiment has an effect, **other than** what you're testing.

E.g. in Experiment C, the **negative control** would be all the equipment set up as normal but **without** the pondweed. If **no oxygen** was produced at any temperature it would show that the variation in the volume of oxygen produced when there was pondweed was due to the **effect** of temperature on the pondweed, and **not** the effect of temperature on **anything else** in the experiment.

6) **Repetition by other scientists** — for theories to become accepted as 'fact' other scientists need to **repeat** the work (see page 2). If **multiple studies** or **experiments** come to the same conclusion, then that conclusion is **more reliable**.

E.g. if a second group of scientists repeated Study B and got the same results, the results would be **more reliable**.

*There Are a Few **Technical Terms** You **Need to Understand***

I'm sure you probably know these all off by heart, but it's easy to get mixed up sometimes. So here's a quick recap of some words **commonly used** when assessing and analysing experiments and studies:

1) **Variable** — A variable is a **quantity** that has the **potential to change**, e.g. weight. There are two types of variable commonly referred to in experiments:
   - **Independent variable** — the thing that's **changed** in an experiment.
   - **Dependent variable** — the thing that you **measure** in an experiment.

When drawing graphs, the dependent variable should go on the **y-axis** (the vertical axis) and the independent on the **x-axis** (the horizontal axis).

2) **Accurate** — Accurate results are those that are **really close** to the **true** answer. The true answer is **without error**, so if you can reduce error as much as possible you'll get a more accurate result. The most **accurate methods** are those that produce as **error-free** results as possible.

3) **Precise results** — These are results taken using **sensitive instruments** that measure in **small increments**, e.g. pH measured with a meter (pH 7.692) will be **more precise** than pH measured with paper (pH 8).

It's possible for results to be precise **but not** accurate, e.g. a balance that weighs to 1/1000 th of a gram will give precise results, but if it's not **calibrated** properly the results won't be accurate.

4) **Qualitative** — A **qualitative** test tells you **what's** present, e.g. an acid or an alkali.

5) **Quantitative** — A **quantitative** test tells you **how much** is present, e.g. an acid that's pH 2.46.

There's enough evidence here to conclude that data interpretation is boring...

*These pages should give you a fair idea of how to interpret data. Just use your head and remember the four things you might be asked to do — **d**escribe the **d**ata, **c**heck the **c**onclusions, **e**xplain the **e**vidence and check the **r**esults are **r**eliable.*

Answers

Unit 4: Section 1 — Communication and Homeostasis

Page 5 — Communication and Homeostasis Basics

1 a) *Maximum of 1 mark available.*
The maintenance of a constant internal environment ***[1 mark]****.*

b) *Maximum of 3 marks available.*
Receptors detect when a level is too high or too low ***[1 mark]****, and the information's communicated via the nervous system or the hormonal system to effectors* ***[1 mark]****. Effectors respond to counteract the change / to bring the level back to normal* ***[1 mark]****.*

Page 7 — The Nervous System and Neurones

1 *Maximum of 5 marks available.*
Receptors detect the stimulus ***[1 mark]****, e.g. light receptors/photoreceptors in the animal's eyes detect the bright light* ***[1 mark]****. The receptors send impulses along neurones via the CNS to the effectors* ***[1 mark]****. The effectors bring about a response* ***[1 mark]****, e.g. the circular iris muscles contract to constrict the pupils and protect the eyes* ***[1 mark]****.*

2 *Maximum of 4 marks available.*
A — receptor cell ***[1 mark]***
B — dendron ***[1 mark]***
C — cell body ***[1 mark]***
D — axon ***[1 mark]***

Page 9 — Action Potentials

1 a) *Maximum of 1 mark available.*
Stimulus ***[1 mark]****.*

b) *Maximum of 3 marks available.*
A stimulus causes sodium ion channels in the neurone cell membrane to open ***[1 mark]****. Sodium ions diffuse into the cell* ***[1 mark]****, so the membrane becomes depolarised* ***[1 mark]****.*

c) *Maximum of 2 marks available.*
The membrane was in the refractory period ***[1 mark]****, so the sodium ion channels were recovering and couldn't be opened* ***[1 mark]****.*

2 *Maximum of 5 marks available.*
Transmission of action potentials will be slower in neurones with damaged myelin sheaths ***[1 mark]****. This is because myelin is an electrical insulator* ***[1 mark]****, so increases the speed of action potential conduction* ***[1 mark]****. The action potentials 'jump' between the nodes of Ranvier/between the myelin sheaths* ***[1 mark]****, where sodium ion channels are concentrated* ***[1 mark]****.*
Don't panic if a question mentions something you haven't learnt about. You might not know anything about multiple sclerosis but that's fine, because you're not supposed to. All you need to know to get full marks here is how myelination affects the speed of action potential conduction.

Page 11 — Synapses

1 *Maximum of 5 marks available.*
A — presynaptic membrane ***[1 mark]***
B — vesicle/vesicle containing neurotransmitter ***[1 mark]***
C — synaptic cleft ***[1 mark]***
D — postsynaptic receptor ***[1 mark]***
E — postsynaptic membrane ***[1 mark]***

2 *Maximum of 6 marks available, from any of the 10 points below.*
The action potential arriving at the presynaptic membrane stimulates voltage-gated calcium ion channels to open ***[1 mark]****, so calcium ions diffuse into the neurone* ***[1 mark]****. This causes synaptic vesicles, containing neurotransmitter, to move to the presynaptic membrane* ***[1 mark]****. They then fuse with the presynaptic membrane* ***[1 mark]****. The vesicles release the neurotransmitter into the synaptic cleft* ***[1 mark]****. The neurotransmitter diffuses across the synaptic cleft* ***[1 mark]*** *and binds to specific receptors on the postsynaptic membrane* ***[1 mark]****. This causes sodium ion channels in the postsynaptic membrane to open* ***[1 mark]****. The influx of sodium ions causes depolarisation* ***[1 mark]****. This triggers a new action potential to be generated at the postsynaptic membrane* ***[1 mark]****.*

Page 13 — The Hormonal System and Glands

1 *Maximum of 4 marks available.*
The first messenger is a hormone ***[1 mark]****, which carries the message from an endocrine gland to the receptor on its target tissue* ***[1 mark]****. The second messenger is a signalling molecule* ***[1 mark]****, which carries the message from the receptor to other parts of the cell and activates a cascade inside the cell* ***[1 mark]****.*

2 *Maximum of 2 marks available.*
Endocrine glands secrete chemicals directly into the blood, but exocrine glands secrete into ducts ***[1 mark]****.*
Endocrine glands secrete hormones, but exocrine glands usually secrete enzymes ***[1 mark]****.*

Page 15 — Homeostasis — Control of Body Temperature

1 *Maximum of 4 marks available.*
Snakes are ectotherms ***[1 mark]****. They can't control their body temperature internally and depend on the temperature of their external environment* ***[1 mark]****. In cold climates, snakes will be less active* ***[1 mark]****, which makes it harder to catch prey, avoid predators, find a mate, etc.* ***[1 mark]****.*
You need to use a bit of common sense to answer this question — you know that the activity level of an ectotherm depends on the temperature of the surroundings, so in a cold environment it won't be very active. And if it can't be very active it'll have trouble surviving.

2 *Maximum of 4 marks available, from any of the 8 points below.*
1 mark for each method, up to a maximum of 2 marks. 1 mark for each explanation, up to a maximum of 2 marks.
Vasoconstriction of blood vessels ***[1 mark]*** *reduces heat loss because less blood flows through the capillaries in the surface layers of the dermis* ***[1 mark]****. Erector pili muscles contract to make hairs stand on end* ***[1 mark]****, trapping an insulating layer of air to prevent heat loss* ***[1 mark]****. Muscles contract in spasms to make the body shiver* ***[1 mark]****, so more heat is produced from increased respiration* ***[1 mark]****. Adrenaline and thyroxine are released* ***[1 mark]****, which increase metabolism so more heat is produced* ***[1 mark]****.*

3 *Maximum of 2 marks available.*
Thermoreceptors/temperature receptors in the skin detect a higher external temperature than normal ***[1 mark]****.*
The thermoreceptors/temperature receptors send impulses along sensory neurones to the hypothalamus ***[1 mark]****.*

Page 17 — Homeostasis — Control of Blood Glucose

1 *Maximum of 5 marks available, from any of the 7 points below.*
High blood glucose concentration is detected by cells in the pancreas ***[1 mark]****. Beta/β cells secrete insulin into the blood* ***[1 mark]****, which binds to receptors on the cell membranes of liver and muscle cells* ***[1 mark]****. This increases the permeability of the cell membranes to glucose, so the cells take up more glucose* ***[1 mark]****. Insulin also activates glycogenesis* ***[1 mark]*** *and increases the rate that cells respire glucose* ***[1 mark]****. This lowers the concentration of glucose in the blood* ***[1 mark]****.*
You need to get the spelling of words like glycogenesis right in the exam or you'll miss out on marks.

2 *Maximum of 2 marks available.*
No insulin would be secreted ***[1 mark]*** *because ATP wouldn't be produced, so the potassium ion channels in the β cell plasma membrane wouldn't close / the plasma membrane of β cell wouldn't be depolarised* ***[1 mark]****.*

Page 19 — Diabetes and Control of Heart Rate

1 *Maximum of 3 marks available.*
They have Type II diabetes ***[1 mark]****. They produce insulin, but the insulin receptors on their cell membranes don't work properly, so the cells don't take up enough glucose* ***[1 mark]****. This means their blood glucose concentration remains higher than normal* ***[1 mark]****.*

Answers

2 *Maximum of 2 marks available, from any of the 4 points below.*
It's cheaper to produce insulin using GM bacteria than to extract it from animal pancreases ***[1 mark]****. Large amounts of insulin can be made using GM bacteria, so there's enough insulin to treat everyone with Type I diabetes* ***[1 mark]****. GM bacteria make real human insulin, which is more effective and less likely to trigger an allergic response or be rejected by the immune system* ***[1 mark]****. Some people prefer insulin from GM bacteria for ethical or religious reasons* ***[1 mark]****.*

3 a) *Maximum of 5 marks available.*
High blood pressure is detected by pressure receptors in the aorta called baroreceptors ***[1 mark]****. Impulses are sent along sensory neurones to the medulla* ***[1 mark]****. Impulses are then sent from the medulla to the SAN along the vagus nerve* ***[1 mark]****. The vagus nerve secretes acetylcholine, which binds to receptors on the sinoatrial node/SAN* ***[1 mark]****. This slows the heart rate (reducing blood pressure)* ***[1 mark]****.*

b) *Maximum of 2 marks available.*
No impulses sent from the medulla would reach the SAN ***[1 mark]****, so the heart rate wouldn't increase or decrease/control of the heart rate would be lost* ***[1 mark]****.*

Unit 4: Section 2 — Excretion

Page 21 — The Liver and Excretion

1 *Maximum of 3 marks available.*
A — central vein ***[1 mark]****, B — sinusoid* ***[1 mark]****,*
C — hepatocyte ***[1 mark]***

2 *Maximum of 6 marks available.*
The protein would be digested, producing amino acids ***[1 mark]****. Amino acids contain nitrogen in their amino groups, but the body can't usually store nitrogenous substances, so if a lot of protein is eaten there could be an excess of amino acids that will need to be used or broken down and excreted* ***[1 mark]****. Excess amino acids are broken down in the liver into ammonia and organic acids in a process called deamination* ***[1 mark]****. Ammonia is then combined with* CO_2 *in the ornithine cycle to produce urea* ***[1 mark]****. Urea is then released into the blood and filtered out at the kidneys to produce urine* ***[1 mark]****. So if a large amount of protein is eaten, there may be excess amino acids that are broken down by the liver, producing a large amount of urea that's excreted in the urine* ***[1 mark]****.*
Don't forget to say that only <u>excess</u> amino acids are broken down.

Page 23 — The Kidneys and Excretion

1 a) *Maximum of 4 marks available.*
A — nephron ***[1 mark]***
B — renal capsule / Bowman's capsule ***[1 mark]***
C — proximal convoluted tubule / PCT ***[1 mark]***
D — collecting duct ***[1 mark]***

b) *Maximum of 1 mark available.*
B (renal capsule) ***[1 mark]***

c) *Maximum of 5 marks available.*
Ultrafiltration is when substances are filtered out of the blood and enter the tubules in the kidneys ***[1 mark]****. Blood enters a glomerulus, a bundle of capillaries looped inside a hollow ball called a renal capsule/Bowman's capsule* ***[1 mark]****. The blood in the glomerulus is under high pressure because it enters through the afferent arteriole and leaves through the smaller efferent arteriole* ***[1 mark]****. The high pressure forces liquid and small molecules in the blood out of the capillary and into the renal capsule* ***[1 mark]****. The liquid and small molecules pass through the capillary wall, the basement membrane and slits in the epithelium of the renal capsule. But larger molecules like proteins and blood cells can't pass through and stay in the blood* ***[1 mark]****.*

Page 25 — Controlling Water Content

1 *Maximum of 6 marks available.*
Near the top of the ascending limb of the loop of Henle, sodium/Na^+ *and chloride/*Cl^- *ions are actively pumped out into the medulla. This creates a low water potential in the medulla* ***[1 mark]****. There's now a lower water potential in the medulla than in the descending limb* ***[1 mark]****, so water moves out of the descending limb and into the medulla by osmosis* ***[1 mark]****. Near the bottom of the ascending limb sodium/*Na^+ *and chloride/*Cl^- *ions diffuse into the medulla, lowering the water potential of the medulla further* ***[1 mark]****. The low water potential in the medulla causes water to move out of the collecting duct by osmosis* ***[1 mark]****. The water in the medulla is then reabsorbed into the blood through the capillary network* ***[1 mark]****.*

2 *Maximum of 6 marks available.*
Strenuous exercise causes more sweating, so more water is lost ***[1 mark]****. This decreases the water content of the blood, so its water potential drops* ***[1 mark]****. This is detected by osmoreceptors in the hypothalamus* ***[1 mark]****, which stimulates the posterior pituitary gland to release more ADH* ***[1 mark]****.*
The answer up to this point has explained the <u>cause</u> of the increase in ADH in the blood. After this, the answer explains the <u>effect</u> on the kidney.
The ADH increases the permeability of the walls of the distal convoluted tubule and collecting duct ***[1 mark]****. This means more water is reabsorbed into the medulla and into the blood by osmosis, so a small amount of concentrated urine is produced* ***[1 mark]****.*

Page 27 — Kidney Failure and Detecting Hormones

1 *Maximum of 5 marks available. For full marks answers must include at least 1 advantage and 1 disadvantage.*
Kidney transplants are cheaper in the long term than renal dialysis ***[1 mark]****. Having a kidney transplant is more convenient for a person than regular dialysis sessions* ***[1 mark]****. A patient who has had a kidney transplant won't feel unwell between dialysis sessions* ***[1 mark]****. However, a transplant means the patient has to undergo a major operation, which is risky* ***[1 mark]****. The patient also has to take drugs to suppress the immune system, so it doesn't reject the transplant* ***[1 mark]****.*

2 *Maximum of 5 marks available.*
Steroids are removed from the blood in the urine, so urine can be tested to see if a person is using steroids ***[1 mark]****. It's tested using a technique called gas chromatography, where the urine is vaporised and passed through a column containing a liquid* ***[1 mark]****. Different substances move through the column at different speeds* ***[1 mark]****. The time taken for substances in the sample to pass through the column is compared to the time taken for a steroid to pass through the column* ***[1 mark]****. If the time taken is the same then the sample contains the steroid* ***[1 mark]****.*

Unit 4: Section 3 — Photosynthesis and Respiration

Page 29 — Photosynthesis, Respiration and ATP

1 *Maximum of 6 marks available, from any of the 8 points below.*
In the cell, ATP is synthesised from ADP and inorganic phosphate/P_i ***[1 mark]*** *using energy from an energy-releasing reaction, e.g. respiration* ***[1 mark]****. The energy is stored as chemical energy in the phosphate bond* ***[1 mark]****. ATP synthase catalyses this reaction* ***[1 mark]****. ATP then diffuses to the part of the cell that needs energy* ***[1 mark]****. Here, it's broken down back into ADP and inorganic phosphate/*P_i ***[1 mark]****, which is catalysed by ATPase* ***[1 mark]****. Chemical energy is released from the phosphate bond and used by the cell* ***[1 mark]****.*
Make sure you don't get the two enzymes confused
— ATP **<u>syn</u>**thase **<u>syn</u>**thesises ATP, and ATPase breaks it down.

Answers

Page 33 — Photosynthesis

1 a) *Maximum of 1 mark available.*
The thylakoid membranes ***[1 mark]****.*

b) *Maximum of 1 mark available.*
Photosystem II ***[1 mark]****.*

c) *Maximum of 4 marks available.*
Light energy splits water ***[1 mark]****.*

H_2O ***[1 mark]*** $\rightarrow 2H^+ + \frac{1}{2}O_2$ ***[1 mark]****.*

The electrons from the water replace the electrons lost from chlorophyll ***[1 mark]****.*

The question asks you to explain the purpose of photolysis, so make sure you include why the water is split up — to replace the electrons lost from chlorophyll.

d) *Maximum of 1 mark available.*
NADP ***[1 mark]****.*

2 a) *Maximum of 6 marks available.*
Ribulose bisphosphate/RuBP and carbon dioxide/CO_2 join together to form an unstable 6-carbon compound ***[1 mark]****. This reaction is catalysed by the enzyme rubisco/ribulose bisphosphate carboxylase* ***[1 mark]****. The compound breaks down into two molecules of a 3-carbon compound called glycerate 3-phosphate/GP* ***[1 mark]****. Two molecules of glycerate 3-phosphate are then converted into two molecules of triose phosphate/TP* ***[1 mark]****. The energy for this reaction comes from ATP* ***[1 mark]*** *and the H^+ ions come from reduced NADP* ***[1 mark]****.*

b) *Maximum of 2 marks available.*
Ribulose bisphosphate is regenerated from triose phosphate/TP molecules ***[1 mark]****. ATP provides the energy to do this* ***[1 mark]****.*

This question is only worth two marks so only the main facts are needed, without the detail of the number of molecules.

c) *Maximum of 3 marks available.*
No glycerate 3-phosphate/GP would be produced ***[1 mark]****, so no triose phosphate/TP would be produced* ***[1 mark]****. This means there would be no glucose produced* ***[1 mark]****.*

Page 35 — Limiting Factors in Photosynthesis

1 *Maximum of 4 marks available.*
25 °C ***[1 mark]****. This is because photosynthesis involves enzymes* ***[1 mark]****, which become inactive at low temperatures/10 °C* ***[1 mark]*** *and denature at high temperatures/45 °C* ***[1 mark]****.*

Page 37 — Limiting Factors in Photosynthesis

1 a) *Maximum of 3 marks available.*
The level of GP will rise and levels of TP and RuBP will fall ***[1 mark]****. This is because there's less reduced NADP and ATP from the light-dependent reaction* ***[1 mark]****, so the conversion of GP to TP and RuBP is slow* ***[1 mark]****.*

b) *Maximum of 3 marks available.*
The levels of RuBP, GP and TP will fall ***[1 mark]****. This is because the reactions in the Calvin cycle are slower* ***[1 mark]*** *due to all the enzymes working more slowly* ***[1 mark]****.*

2 *Maximum of 6 marks available, from any of the 8 points below.*
A sample of pondweed would be placed in a test tube of water ***[1 mark]****. The test tube would be placed in a beaker containing water at a known temperature* ***[1 mark]****. The test tube would be connected to a capillary tube of water* ***[1 mark]*** *and the capillary tube connected to a syringe* ***[1 mark]****. The pondweed would be allowed to photosynthesise for a set period of time* ***[1 mark]****. Afterwards, the syringe would be used to draw the bubble of oxygen produced up the capillary tube where its length would be measured using a ruler* ***[1 mark]****. The experiment is repeated and the mean length of gas bubble is calculated* ***[1 mark]****. Then the whole experiment is repeated at several different temperatures* ***[1 mark]****.*

Page 39 — Aerobic Respiration

1 *Maximum of 6 marks available, from any of the 8 points below.*
First, the 6-carbon glucose molecule is phosphorylated ***[1 mark]*** *by adding two phosphates from two molecules of ATP* ***[1 mark]****. This creates one molecule of 6-carbon hexose bisphosphate* ***[1 mark]*** *and two molecules of ADP* ***[1 mark]****. Then, the hexose bisphosphate is split up into two molecules of 3-carbon triose phosphate* ***[1 mark]****. Triose phosphate is oxidised (by removing hydrogen) to give two molecules of 3-carbon pyruvate* ***[1 mark]****. The hydrogen is accepted by two molecules of NAD, producing two molecules of reduced NAD* ***[1 mark]****. During oxidation four molecules of ATP are produced* ***[1 mark]****.*

When describing glycolysis make sure you get the number of molecules correct — one glucose molecule produces one molecule of hexose bisphosphate, which produces two molecules of triose phosphate. You could draw a diagram in the exam to show the reactions.

2 a) *Maximum of 3 marks available, from any of the 4 points below.*
The 3-carbon pyruvate is decarboxylated ***[1 mark]*** *and NAD is reduced to form acetate* ***[1 mark]****. Acetate combines with coenzyme A (CoA) to form acetyl coenzyme A (acetyl CoA)* ***[1 mark]****. No ATP is produced* ***[1 mark]****.*

b) *Maximum of 2 marks available, from any of the 3 points below.*
The inner membrane is folded into cristae, which increase the membrane's surface area and maximise respiration ***[1 mark]****. There are lots of ATP synthase molecules on the inner membrane to produce lots of ATP in the final stage of respiration* ***[1 mark]****. The matrix contains all the reactants and enzymes needed for the Krebs cycle to take place* ***[1 mark]****.*

Page 41 — Aerobic Respiration

1 a) *Maximum of 2 mark available.*
The transfer of electrons down the electron transport chain stops ***[1 mark]****. So there's no energy released to phosphorylate ADP/produce ATP* ***[1 mark]****.*

b) *Maximum of 2 marks available.*
The Krebs cycle stops ***[1 mark]*** *because there's no oxidised NAD/FAD coming from the electron transport chain* ***[1 mark]****.*

Part b is a bit tricky — remember that when the electron transport chain is inhibited, the reactions that depend on the products of the chain are also affected.

Page 43 — Respiration Experiments

1 a) *Maximum of 1 mark available.*
Because there was no proton gradient ***[1 mark]****.*

b) *Maximum of 1 mark available.*
3.7 ***[1 mark]***

c) *Maximum of 1 mark available.*
Yes, these results support the chemiosmotic theory because they show that a proton gradient can be used by mitochondria to synthesise ATP ***[1 mark]****.*

2 a) *Maximum of 1 mark available.*
To make sure the results are only due to oxygen uptake by the woodlouse ***[1 mark]****.*

b) *Maximum of 2 marks available.*
The oxygen taken up would be replaced by carbon dioxide given out / there would be no change in air volume in the test tube ***[1 mark]****. This means there would be no movement of the liquid in the manometer* ***[1 mark]****.*

c) *Maximum of 1 mark available.*
Carbon dioxide/CO_2 ***[1 mark]****.*

Page 45 — Aerobic and Anaerobic Respiration

1 *Maximum of 1 mark available.*
Because lactate fermentation doesn't involve electron carriers/the electron transport chain/oxidative phosphorylation ***[1 mark]****.*

2 *Maximum of 2 marks available.*
$RQ = CO_2 \div O_2$ ***[1 mark]***
So the RQ of triolein = 57 ÷ 80 = 0.71 ***[1 mark]***
Award 2 marks for the correct answer of 0.71, without any working.

Answers

Unit 5: Section 1 — Protein Synthesis and Cellular Control

Page 47 — DNA, RNA and Protein Synthesis

1 *Maximum of 2 marks available.*
*mRNA carries the genetic code from the DNA in the nucleus to the cytoplasm, where it's used to make a protein during translation **[1 mark]**. tRNA carries the amino acids that are used to make proteins to the ribosomes during translation **[1 mark]**.*

2 a) *Maximum of 1 mark available.*
*5 amino acids **[1 mark]***

b) *Maximum of 2 marks available. Award 2 marks if all five amino acids are correct and in the correct order. Award 1 mark if four amino acids are correct and in the correct order.*
AGA = serine
ATA = tyrosine
CAC = valine
CGT = alanine
Correct sequence = serine, serine, tyrosine, valine, alanine.

Page 49 — Transcription and Translation

1 *Maximum of 2 marks available.*
*The drug binds to DNA, preventing RNA polymerase from binding, so transcription can't take place and no mRNA can be made **[1 mark]**. This means there's no mRNA for translation and so protein synthesis is inhibited **[1 mark]**.*

2 a) *Maximum of 2 marks available.*
*10 × 3 = 30 nucleotides long **[1 mark]**. Each amino acid is coded for by three nucleotides (a codon), so the mRNA length in nucleotides is the number of amino acids multiplied by three **[1 mark]**.*

b) *Maximum of 6 marks available.*
*The mRNA attaches itself to a ribosome and transfer RNA (tRNA) molecules carry amino acids to the ribosome **[1 mark]**. A tRNA molecule, with an anticodon that's complementary to the first codon on the mRNA (the start codon), attaches itself to the mRNA by complementary base pairing **[1 mark]**. A second tRNA molecule attaches itself to the next codon on the mRNA in the same way **[1 mark]**. The two amino acids attached to the tRNA molecules are joined by a peptide bond and the first tRNA molecule moves away, leaving its amino acid behind **[1 mark]**. A third tRNA molecule binds to the next codon on the mRNA and its amino acid binds to the first two and the second tRNA molecule moves away **[1 mark]**. This process continues, producing a chain of linked amino acids (a polypeptide chain), until there's a stop codon on the mRNA molecule **[1 mark]**.*

Page 51 — Control of Protein Synthesis and Body Plans

1 *Maximum of 4 marks available.*
*When no lactose is present, the lac repressor binds to the operator site and blocks transcription **[1 mark]**. When lactose is present, it binds to the lac repressor **[1 mark]**, changing its shape so that it can no longer bind to the operator site **[1 mark]**. RNA polymerase can now begin transcription of the structural genes, including the ones that code for β-galactosidase and lactose permease **[1 mark]**.*

Page 53 — Protein Activation and Gene Mutation

1 a) *Maximum of 1 mark available.*
*Mutations are changes to the base sequence/nucleotide sequence of DNA **[1 mark]**.*

b) *Maximum of 2 marks available, from any of the 5 points below.*
*Substitution — one base is swapped for another **[1 mark]**.*
*Deletion — one base is removed **[1 mark]**.*
*Insertion — one base is added **[1 mark]**.*
*Duplication — one or more bases are repeated **[1 mark]**.*
*Inversion — a sequence of bases is reversed **[1 mark]**.*

2 a) *Maximum of 1 mark available.*
*ATGTATTCCGCTGT **[1 mark]***

b) *Maximum of 3 marks available.*
*The mutation changes a triplet in the gene from TCA to TCC **[1 mark]**. But the mutated triplet still codes for serine **[1 mark]**, so the mutation would have a neutral effect on the protein that the gene codes for **[1 mark]**.*

Unit 5: Section 2 — Inheritance

Page 55 — Meiosis

1 a) *Maximum of 4 marks available.*
*The chromosomes condense, getting shorter and fatter **[1 mark]**. Homologous chromosomes pair up **[1 mark]**. The centrioles start moving to opposite ends of the cell, forming a network of protein fibres across it called the spindle **[1 mark]**. The nuclear envelope breaks down **[1 mark]**.*
The question asks you to describe the nuclear envelope, chromosomes and centrioles, so make sure you include them all to get full marks.

b) *Maximum of 2 marks available.*
*A — Telophase II **[1 mark]***
*B — Anaphase II **[1 mark]***

2 a) *Maximum of 7 marks available.*
*Crossing-over of chromatids during prophase I causes genetic variation **[1 mark]**. The non-sister chromatids twist around each other and bits of the chromatids swap over **[1 mark]**. This means that each of the four daughter cells contain chromatids with different combinations of alleles **[1 mark]**. Independent assortment of chromosomes in metaphase I produces genetic variation **[1 mark]**. Different combinations of maternal and paternal chromosomes go into each daughter cell, so each cell ends up with a different combination of alleles **[1 mark]**. Independent assortment of chromatids in metaphase II also produces genetic variation **[1 mark]**. Different combinations of chromatids go into each daughter cell, so each cell ends up with a different combination of alleles **[1 mark]**.*

b) *Maximum of 1 mark available.*
*Fertilisation increases genetic variation because any egg cell can fuse with any sperm cell **[1 mark]**.*

Page 57 — Inheritance

1 a) *Maximum of 3 marks available.*
*Parents' genotypes identified as X^HX^h and X^hY **[1 mark]**.*
*Correct genetic diagram drawn with gametes' alleles identified as X^H, X^h and X^h,Y **[1 mark]** and gametes crossed to show X^HX^h, X^HY, X^hX^h and X^hY as the possible genotypes **[1 mark]**.*
The question specifically asks you to draw a genetic diagram, so make sure that you include one in your answer, e.g.

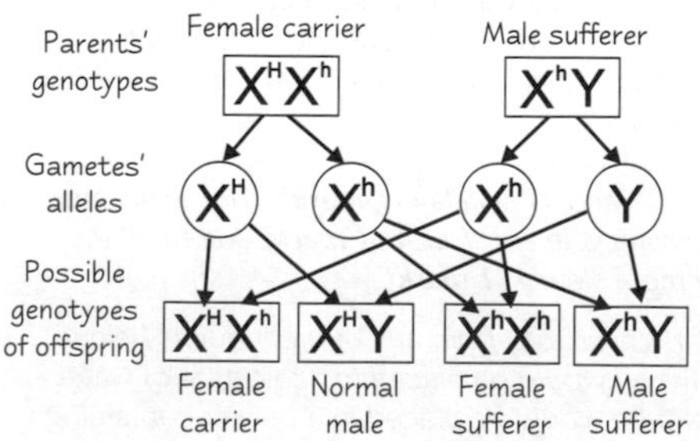

b) *Maximum of 3 marks available.*
*Men only have one copy of the X chromosome (XY) but women have two (XX) **[1 mark]**. Haemophilia A is caused by a recessive allele so females would need two copies of the allele for them to have haemophilia A **[1 mark]**. As males only have one X chromosome they only need one recessive allele to have haemophilia A, which makes them more likely to have haemophilia A than females **[1 mark]**.*

Answers

Page 59 — Phenotypic Ratios and Epistasis

1 *Maximum of 3 marks available.*
Parents' genotypes identified as RRgg and rrGG ***[1 mark]****.*
Correct genetic diagram drawn with gametes' alleles identified as Rg and rG ***[1 mark]*** *and gametes crossed to show RrGg as the only possible genotype of the offspring* ***[1 mark]****.*
The question specifically asks you to draw a genetic diagram, so make sure that you include one in your answer, e.g.

Parents' alleles

Gametes' alleles

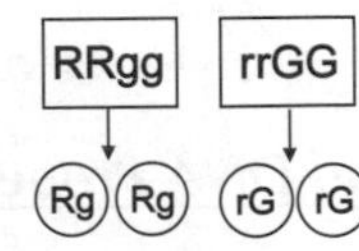

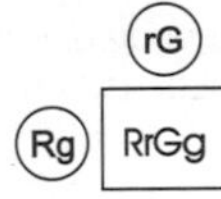

2 *Maximum of 4 marks available.*
A cross between CCGG and ccgg will produce a 9 : 3 : 4 phenotypic ratio in the F_2 generation ***[1 mark]*** *of coloured grey : coloured black : albino* ***[1 mark]****. This is because gene 1 has a recessive epistatic gene (c)* ***[1 mark]****, and two copies of the recessive epistatic gene (cc) will mask the expression of the colour gene* ***[1 mark]****.*
You don't need to draw a genetic diagram to explain the phenotypic ratio that you'd expect from this cross. You can just state the ratio and explain it using your own knowledge.

3 *Maximum of 3 marks available.*
The table shows that a cross between hhss and HHSS produces a 36 : 9 : 3 or 12 : 3 : 1 phenotypic ratio in the F_2 generation of bald : straight hair : curly hair ***[1 mark]****. This is because the hair gene has a dominant epistatic allele (H)* ***[1 mark]****, which means having at least one copy of the dominant epistatic gene (Hh or HH) will result in a bald phenotype that masks the expression of the type of hair gene* ***[1 mark]****.*

Page 61 — The Chi-Squared Test

1 a) *Maximum of 4 marks available.*
(1 mark for each correct column and 1 mark for the answer).

Phenotype	Ratio	Expected Result (E)	Observed Result (O)	O – E	O – E²	(O – E²)/E
Blue with white spots	9	135	131	–4	16	0.12
Purple with white spots	3	45	52	7	49	1.09
Blue with yellow spots	3	45	48	3	9	0.2
Purple with yellow spots	1	15	9	–6	36	2.4
						3.81

b) *Maximum of 2 marks available.*
The χ^2 value does support the null hypothesis ***[1 mark]*** *because it's smaller than the critical value* ***[1 mark]****.*

Unit 5: Section 3 — Variation and Evolution

Page 63 — Variation

1 a) *Maximum of 1 mark available.*
Discontinuous variation ***[1 mark]****.*
b) *Maximum of 1 mark available.*
18.99 – 9.25 = 9.74 kg ***[1 mark]***
c) *Maximum of 2 marks available.*
Mass ***[1 mark]*** *because it shows continuous variation* ***[1 mark]****.*

2 *Maximum of 2 marks available.*
E.g. body mass ***[1 mark]*** *because large parents often have large children so it's affected by genotype, but body mass is also influenced by diet and exercise, which are environmental factors* ***[1 mark]****.*

Page 65 — Evolution by Natural Selection and Genetic Drift

1 *Maximum of 5 marks available.*
The dark moths were better camouflaged on the blackened trees than the pale moths ***[1 mark]****. This means the dark moths were better adapted to avoid predation than the pale moths* ***[1 mark]****. Moths with the allele for a darker colour were more likely to survive, reproduce and pass on their alleles than pale moths* ***[1 mark]****. So a greater proportion of the next generation inherited the allele for a darker colour* ***[1 mark]****. The frequency of this allele then increased from generation to generation, causing an increase in the number of dark moths* ***[1 mark]****.*

Page 67 — Hardy-Weinberg Principle and Artificial Selection

1 *Maximum of 1 mark available.*
Frequency of the recessive allele (q) = 0.23, and p + q = 1
So the frequency of the dominant allele (p) = 1 – q
= 1 – 0.23
= 0.77 ***[1 mark]****.*

2 *Maximum of 3 marks available.*
Farmers could have selected a male and female with a high meat yield and bred these two together ***[1 mark]****. Then they could have selected the offspring with the highest meat yields and bred them together* ***[1 mark]****. This process could have been continued over several generations to produce cattle with a very high meat yield* ***[1 mark]****.*

Page 69 — Speciation

1 a) *Maximum of 1 mark available.*
The new species could not breed with each other ***[1 mark]****.*
b) *Maximum of 3 marks available.*
Different populations of flies were isolated and fed on different foods ***[1 mark]****. This caused changes in allele frequencies between the populations* ***[1 mark]****, which made them reproductively isolated and eventually resulted in speciation* ***[1 mark]****.*
c) *Maximum of 2 marks available, from any of the 3 points below.*
Seasonal changes (become sexually active at different times) ***[1 mark]****. Mechanical changes (changes to genitalia)* ***[1 mark]****. Behavioural changes (changes in behaviour that prevent mating)* ***[1 mark]****.*
d) *Maximum of 1 mark available, from any of the 5 points below or any other good point.*
E.g. geographical barrier ***[1 mark]****, flood* ***[1 mark]****, volcanic eruption* ***[1 mark]****, earthquake* ***[1 mark]****, glacier* ***[1 mark]****.*

Unit 5: Section 4 — Cloning and Biotechnology

Page 72 — Cloning

1 a) *Maximum of 2 marks available.*
Reproductive cloning is used to make a complete organism that's genetically identical to another organism ***[1 mark]****. Non-reproductive cloning is used to make embryonic stem cells that are genetically identical to another organism* ***[1 mark]****.*
b) *Maximum of 6 marks available.*
The scientists could use nuclear transfer ***[1 mark]****. They would take a body cell from an organism (organism A) and extract its nucleus* ***[1 mark]****. An egg cell would be taken from another organism (organism B) and its nucleus would be removed, forming an enucleated egg cell* ***[1 mark]****. The scientists would transfer the body cell nucleus into the enucleated egg cell* ***[1 mark]****. They would then stimulate the egg cell to divide* ***[1 mark]****. An embryo would form, which would be made up of stem cells that are genetically identical to the cells found in organism A* ***[1 mark]****.*
Don't forget — the technique of nuclear transfer is used in both reproductive and non-reproductive cloning.

Answers

Page 75 — Biotechnology

1 Maximum of 8 marks available.
The first phase of the standard growth curve is the lag phase, when the microorganism population increases slowly **[1 mark]**. This is because the microorganisms need to make enzymes and other molecules before they can reproduce **[1 mark]**. The culture then enters the exponential phase, when the population size increases quickly **[1 mark]**. This is because there's lots of food and little competition **[1 mark]**. The next phase is the stationary phase, when the population size stays level **[1 mark]**. This is because the reproductive rate equals the death rate **[1 mark]**. The culture then enters the decline phase, when the population size begins to fall **[1 mark]**. This is because food is scarce and waste products are at toxic levels, causing microorganisms to die **[1 mark]**.

Unit 5: Section 5 — Gene Technologies

Page 77 — Common Techniques

1 Maximum of 6 marks available.
The DNA sample is mixed with free nucleotides, primers and DNA polymerase **[1 mark]**. The mixture is heated to 95 °C to break the hydrogen bonds **[1 mark]**. The mixture is then cooled to between 50 – 65 °C to allow the primers to bind/anneal to the DNA **[1 mark]**. The primers bind/anneal to the DNA because they have a sequence that's complementary to the sequence at the start of the DNA fragment **[1 mark]**. The mixture is then heated to 72 °C and DNA polymerase lines up free nucleotides along each template strand, producing new strands of DNA **[1 mark]**. The cycle would be repeated over and over to produce lots of copies **[1 mark]**.
This question asks you to describe and explain, so you need to give the reasons why each stage is done to gain full marks.

2 Maximum of 5 marks available.
A fluorescent tag is added to all the DNA fragments in the mixture so they can be viewed under UV light **[1 mark]**. The DNA mixture is placed into a well in a slab of gel and covered in a buffer solution that conducts electricity **[1 mark]**. An electrical current is passed through the gel and the DNA fragments move towards the positive electrode because DNA fragments are negatively charged **[1 mark]**. The DNA fragments separate according to size because the small fragments move faster and travel further through the gel **[1 mark]**. The DNA fragments are viewed as bands under UV light **[1 mark]**.

Page 79 — Genetic Engineering

1 a) Maximum of 2 marks available.
Colony A has grown on the agar plate containing penicillin **[1 mark]** so it contains the penicillin-resistance marker gene, which means it contains transformed cells **[1 mark]**.

b) Maximum of 3 marks available.
The plasmid vector DNA would have been cut open with the same restriction endonuclease that was used to isolate the DNA fragment containing the desired gene **[1 mark]**. The plasmid DNA and gene (DNA fragment) would have been mixed together with DNA ligase **[1 mark]**. DNA ligase joins the sugar-phosphate backbone of the two bits of DNA **[1 mark]**.

c) Maximum of 2 marks available.
It's useful for bacteria to take up plasmids because the plasmids may contain useful genes **[1 mark]** that increase their chance of survival **[1 mark]**.

Page 81 — Genetic Engineering

1 Maximum of 6 marks available.
The gene for human insulin is identified and isolated using restriction enzymes **[1 mark]**. A plasmid is cut open using the same restriction enzymes that were used to isolate the insulin gene **[1 mark]**. The insulin gene is inserted into the plasmid **[1 mark]**. The plasmid is taken up by bacteria and any transformed bacteria are identified using marker genes **[1 mark]**. The bacteria are grown in a fermenter and insulin is produced by the bacteria as they grow and divide **[1 mark]**. The insulin is extracted and purified so it can be used in humans **[1 mark]**.

2 Maximum of 7 marks available.
The psy and crtl genes are isolated using restriction enzymes **[1 mark]**. A plasmid is removed from the Agrobacterium tumefaciens bacterium and cut open using the same restriction enzymes **[1 mark]**. The psy and crtl genes and a marker gene are inserted into the plasmid **[1 mark]**. The recombinant plasmid is put back into the A. tumefaciens bacterium **[1 mark]**. Rice plant cells are incubated with the transformed A. tumefaciens bacteria, which infect the rice plant cells **[1 mark]**. A. tumefaciens inserts the genes into the plant cells' DNA **[1 mark]**. The rice plant cells are then grown on a selective medium, so only the transformed rice plants will be able to grow **[1 mark]**.

Page 83 — Gene Therapy and DNA Probes

1 a) Maximum of 1 mark available.
Gene therapy involves altering/supplementing defective genes (mutated alleles) inside cells to treat genetic disorders and cancer **[1 mark]**.

b) Maximum of 1 mark available.
Somatic gene therapy **[1 mark]**.

2 Maximum of 3 marks available, from any 6 of the points below.
E.g. the effect of the treatment may be short-lived **[1 mark]**.
The patient might have to undergo multiple treatments **[1 mark]**.
It might be difficult to get the allele into specific body cells **[1 mark]**.
The body may start an immune response against the vector **[1 mark]**.
The allele may be inserted into the wrong place in the DNA, which could cause more problems **[1 mark]**. The allele may be overexpressed **[1 mark]**.

3 Maximum of 4 marks available.
The separated DNA fragments are transferred to a nylon membrane and incubated with a fluorescently labelled DNA probe **[1 mark]**.
The probe is complementary to the sequence of the mutated BRCA1 gene **[1 mark]**. If the sequence is present in one of the DNA fragments, the DNA probe will hybridise to it **[1 mark]**. The membrane is then exposed to UV light and if the sequence is present in one of the DNA fragments, then that band will fluoresce **[1 mark]**.

Page 85 — Sequencing Genes and Genomes

1 a) Maximum of 3 marks available.
DNA primer **[1 mark]**, free nucleotides **[1 mark]** and fluorescently-labelled modified nucleotides **[1 mark]**.

b) Maximum of 6 marks available.
The reaction mixture is added to four tubes, with a different modified nucleotide in each tube **[1 mark]**. The tubes undergo PCR to produce lots of strands of DNA of different lengths **[1 mark]**. Each strand of DNA is a different length because each one terminates at a different point depending on where the modified nucleotide was added **[1 mark]**. The DNA fragments in each tube are separated by electrophoresis and visualised under UV light **[1 mark]**.
The smallest nucleotide is at the bottom of the gel and each band after this represents one more base added **[1 mark]**.
So the bands can be read from the bottom of the gel to the top, forming the base sequence of the DNA fragment **[1 mark]**.

2 Maximum of 8 marks available.
The genome is cut up into smaller fragments using restriction enzymes **[1 mark]**. The individual fragments are inserted into bacterial artificial chromosomes/BACs, which are then inserted into bacteria **[1 mark]**. Each BAC contains a different DNA fragment, so each bacterium contains a BAC with a different DNA fragment **[1 mark]**. The bacteria divide, creating colonies of cloned cells that contain their specific DNA fragment **[1 mark]**. Together the different colonies make a complete genomic DNA library **[1 mark]**. DNA is extracted from each colony and cut up using restriction enzymes, producing overlapping pieces of DNA **[1 mark]**. Each piece of DNA is sequenced, using the chain-termination method, and the pieces are put back in order to give the full sequence from that BAC **[1 mark]**. Finally the DNA fragment from each different BAC is put back in order, using computers, to complete the entire genome **[1 mark]**.

Answers

Unit 5: Section 6 — Ecology

Page 87 — Ecosystems and the Nitrogen Cycle

1 a) *Maximum of 2 marks available.*
*A — ammonification **[1 mark]**, C — denitrification **[1 mark]***
b) *Maximum of 3 marks available.*
*Process B is nitrogen fixation **[1 mark]**. Nitrogen fixation is where nitrogen gas in the atmosphere is turned into ammonia **[1 mark]** by bacteria **[1 mark]**.*

Page 89 — Energy Transfer Through an Ecosystem

1 a) *Maximum of 4 marks available.*
*Because not all of the energy available from the grass is taken in by the Arctic hare **[1 mark]**. Some parts of the grass aren't eaten, so the energy isn't taken in **[1 mark]**, and some parts of the grass are indigestible, so they'll pass through the hares and come out as waste **[1 mark]**. Some energy is lost to the environment when the Arctic hare uses energy from respiration for things like movement or body heat **[1 mark]**.*
b) *Maximum of 2 marks available.*
*$(137 \div 2345) \times 100 = 5.8$ **[1 mark]***
*Efficiency of energy transfer = 5.8% **[1 mark]***
Award 2 marks for correct answer of 5.8% without any working.

Page 91 — Succession

1 a) *Maximum of 6 marks available.*
*This is an example of secondary succession, because there is already a soil layer present in the field **[1 mark]**. The first species to grow will be the pioneer species, which in this case will be larger plants **[1 mark]**. These will then be replaced with shrubs and smaller trees **[1 mark]**. At each stage, different plants and animals that are better adapted for the improved conditions will move in, out-compete the species already there, and become the dominant species **[1 mark]**. As succession goes on, the ecosystem becomes more complex, so species diversity (the number and abundance of different species) increases **[1 mark]**. Eventually large trees will grow, forming the climax community, which is the final seral stage **[1 mark]**.*
b) *Maximum of 2 marks available.*
*Ploughing destroys any plants that were growing **[1 mark]**, so larger plants may start to grow, but they won't have long enough to establish themselves before the field is ploughed again **[1 mark]**.*

Page 93 — Investigating Ecosystems

1 a) *Maximum of 1 mark available.*
*By taking random samples of the population **[1 mark]**.*
b) *Maximum of 3 marks available.*
*Several frame quadrats would be placed on the ground at random locations within the field **[1 mark]**. The percentage of each frame quadrat that's covered by daffodils would be recorded **[1 mark]**. The percentage cover for the whole field could then be estimated by averaging the data collected in all of the frame quadrats **[1 mark]**.*

Page 95 — Factors Affecting Population Size

1 a) *Maximum of 7 marks available.*
*In the first three years, the population of prey increases from 5000 to 30 000. The population of predators increases slightly later (in the first five years), from 4000 to 11 000 **[1 mark]**. This is because there's more food available for the predators **[1 mark]**. The prey population then falls after year three to 3000 just before year 10 **[1 mark]**, because lots are being eaten by the large population of predators **[1 mark]**. Shortly after the prey population falls, the predator population also falls (back to 4000 by just after year 10) **[1 mark]**, because there's less food available **[1 mark]**. The same pattern is repeated in years 10-20 **[1 mark]**.*
b) *Maximum of 4 marks available.*
*The population of prey increased to around 40 000 by year 26 **[1 mark]**. This is because there were fewer predators, so fewer prey were eaten **[1 mark]**. The population then decreased after year 26 to 25 000 by year 30 **[1 mark]**. This could be because of intraspecific competition **[1 mark]**.*

Page 97 — Conservation of Ecosystems

1 *Maximum of 3 marks available, from any of the 4 points below. For full marks, answers must contain at least one economic, one social and one ethical reason.*
Conservation of ecosystems is important for economic reasons because ecosystems provide resources for things that are traded on a local and global scale, like clothes, drugs and food.
*If they're not conserved, the resources could be lost, causing large economic losses in the future **[1 mark]**. Many ecosystems bring joy to lots of people because they're attractive to look at and people use them for activities like birdwatching and walking. If they aren't conserved the ecosystems may be lost, so future generations won't be able to use and enjoy them **[1 mark]**. Some people think ecosystems should be conserved because it's the right thing to do. They think organisms have a right to exist, so they shouldn't become extinct because of human activity **[1 mark]**. Some people also think that humans have a moral responsibility to conserve ecosystems for future human generations, so they can enjoy and use them **[1 mark]**.*

2 a) *Maximum of 2 marks available.*
1 mark for an explanation and 1 mark for an example.
*Non-native animal species eat some native species, causing a decrease in the populations of native species **[1 mark]**. For example, dogs, cats and black rats eat young giant tortoises and Galapagos land iguanas **[1 mark]** / pigs destroy the nests of Galapagos land iguanas and eat their eggs **[1 mark]** / goats have eaten a lot of the plant life on some of the islands **[1 mark]**.*
b) *Maximum of 2 marks available.*
*Non-native plant species have decreased native plant populations because they compete with the native species **[1 mark]**. For example, quinine trees are taller than some native plants. They block out light to the native plants, which then struggle to survive **[1 mark]**.*
c) *Maximum of 2 marks available.*
1 mark for an explanation and 1 mark for an example.
*Fishing has caused a decrease in the populations of some of the sea life around the Galapagos Islands **[1 mark]**. For example, sea cucumber and hammerhead shark populations have been reduced because of overfishing **[1 mark]** / Galapagos green turtle numbers have been reduced because of overfishing **[1 mark]** / Galapagos green turtle numbers have been reduced because they're killed accidentally when they're caught in fishing nets **[1 mark]**.*
You've been asked to explain how specific animals or plants have been affected, so you need to use named examples.

Unit 5: Section 7 — Responding to the Environment

Page 101 — Plant Responses

1 *Maximum of 3 marks available.*
*Auxins are produced in the tip of shoots and they're moved around the plant, so different parts of the plant have different amounts of auxins **[1 mark]**. The uneven distribution of auxins means there's uneven growth of the plant **[1 mark]**. Auxins move to the more shaded parts of the shoots, making the cells there elongate, which makes the shoot bend towards the light **[1 mark]**.*

2 a) *Maximum of 1 mark available.*
*Auxins **[1 mark]**.*
b) *Maximum of 2 marks available.*
*Apical dominance saves energy as it stops side shoots growing. This allows a plant in an area where there are lots of other plants to grow tall very fast, past the smaller plants, to reach the sunlight **[1 mark]**. Apical dominance also prevents side shoots of the same plant from competing with the shoot tip for light **[1 mark]**.*

3 a) *Maximum of 1 mark available.*
*Ethene **[1 mark]**.*
b) *Maximum of 1 mark available.*
*Ethene stimulates enzymes that break down cell walls, break down chlorophyll and convert starch to sugars **[1 mark]**.*
c) *Maximum of 1 mark available.*
*E.g. the tomatoes are less likely to be damaged in transport **[1 mark]**.*

Answers

Page 103 — Animal Responses

1 a) *Maximum of 1 mark available.*
Hypothalamus ***[1 mark]****.*
b) *Maximum of 2 marks available.*
Control of breathing ***[1 mark]****. Control of heart rate* ***[1 mark]****.*
c) *Maximum of 1 mark available.*
Lack of coordinated movement / balance / posture ***[1 mark]****.*
You know that the cerebellum normally coordinates muscles, balance and posture, so damage to it is likely to cause a lack of coordinated movement, balance or posture.

2 a) *Maximum of 1 mark available.*
The 'fight or flight' response is when an organism prepares its body for action, e.g. to fight or run away ***[1 mark]****.*
b) *Maximum of 1 mark available.*
E.g. when an organism is threatened by a predator ***[1 mark]****.*
c) *Maximum of 1 mark available, from any of the 5 points below.*
E.g. heart rate increases ***[1 mark]****. Muscles around the bronchioles relax* ***[1 mark]****. Glycogen is converted into glucose* ***[1 mark]****. Muscles in the arterioles supplying the skin and gut constrict* ***[1 mark]****. Muscles in the arterioles supplying the heart, lungs and skeletal muscles dilate* ***[1 mark]****.*

Page 105 — Muscle Contraction

1 *Maximum of 3 marks available.*
Muscles are made up of bundles of muscle fibres ***[1 mark]****.*
Muscle fibres contain long organelles called myofibrils ***[1 mark]****.*
Myofibrils contain bundles of myofilaments ***[1 mark]****.*

2 a) *Maximum of 3 marks available.*
A = sarcomere ***[1 mark]****, B = Z-line* ***[1 mark]****, C = H-zone* ***[1 mark]****.*
b) *Maximum of 3 marks available.*
Drawing number 3 ***[1 mark]*** *because the M-line connects the middle of the myosin filaments* ***[1 mark]****. The cross-section would only show myosin filaments, which are the thick filaments* ***[1 mark]****.*
The answer isn't drawing number 1 because all the dots in the cross-section are smaller, so the filaments shown are thin actin filaments — which aren't found at the M-line.

Page 107 — Muscle Contraction

1 *Maximum of 2 marks available.*
The A-bands stay the same length during contraction ***[1 mark]****.*
The I-bands get shorter ***[1 mark]****.*

2 *Maximum of 3 marks available.*
Muscles need ATP to relax because ATP provides the energy to break the actin-myosin cross bridges ***[1 mark]****. If the cross bridges can't be broken, the myosin heads will remain attached to the actin filaments* ***[1 mark]****, so the actin filaments can't slide back to their relaxed position* ***[1 mark]****.*

3 *Maximum of 3 marks available.*
The muscles won't contract ***[1 mark]*** *because calcium ions won't be released into the sarcoplasm, so troponin won't be removed from its binding site* ***[1 mark]****. This means no actin-myosin cross bridges can be formed* ***[1 mark]****.*

Page 109 — Muscle Contraction

1 *Maximum of 5 marks available, from any of the 8 points below.*
Both types of muscle have one nucleus per muscle cell/fibre ***[1 mark]****. Both types have cells/fibres that are small/about 0.2 mm long* ***[1 mark]****. Neither type fatigues/gets tired quickly* ***[1 mark]****. Neither type is under conscious control* ***[1 mark]****. However, involuntary muscle is found in the walls of hollow internal organs like the gut, but cardiac muscle is found in the walls of the heart* ***[1 mark]****. Involuntary muscle fibres are spindle-shaped but cardiac muscle fibres are cylinder-shaped with intercalated discs* ***[1 mark]****. Cardiac muscle fibres are branched but involuntary muscles fibres aren't* ***[1 mark]****. Cardiac muscle fibres have some cross-striations but involuntary muscle fibres have a smooth appearance* ***[1 mark]****.*

2 *Maximum of 3 marks available, from any of the 6 points below.*
E.g. The neurotransmitter used at a neuromuscular junction is always acetylcholine, whereas various neurotransmitters can be used at a synapse ***[1 mark]****. The postsynaptic receptors at a neuromuscular junction are always nicotinic cholinergic receptors, whereas they can be various receptors at a synapse depending on the type of neurotransmitter* ***[1 mark]****. Neuromuscular junctions have lots of postsynaptic receptors, whereas synapses have fewer* ***[1 mark]****. The postsynaptic cell at a neuromuscular junction is a muscle cell, whereas at a synapse it's a neurone* ***[1 mark]****. The postsynaptic membrane at a neuromuscular junction has clefts containing acetylcholinesterase/AChE, but the postsynaptic membrane at a synapse is smooth* ***[1 mark]****. At neuromuscular junctions acetylcholine is broken down by acetylcholinesterase/AChE, but at synapses the neurotransmitter is broken down in different ways depending on what it is* ***[1 mark]****.*

Page 111 — Behaviour

1 *Maximum of 2 marks available.*
Classical conditioning has occurred ***[1 mark]****. The postman has learned to respond naturally to the stimulus of approaching Number 10, which wouldn't normally cause that response* ***[1 mark]****.*

2 *Maximum of 2 marks available.*
A dog could be rewarded for good behaviour, e.g. it could be given a biscuit for sitting down when the trainer says, "Sit" ***[1 mark]****. If the dog is repeatedly rewarded for sitting down then that behaviour will be reinforced, and the dog will learn to sit when told* ***[1 mark]****.*
You could answer this question by using an example of punishing a dog for bad behaviour instead.

Page 113 — Behaviour

1 a) *Maximum of 1 mark available.*
Imprinting is where an animal learns to recognise its parents and instinctively follows them ***[1 mark]****.*
b) *Maximum of 2 marks available.*
A gosling can imprint on a human if the gosling is reared from birth/ during the critical period by a human ***[1 mark]****. The human will be the first moving object the gosling sees, so the gosling will imprint on the human/will follow the human* ***[1 mark]****.*

2 a) *Maximum of 1 mark available.*
Social behaviour is behaviour that involves members of a group interacting with each other ***[1 mark]****.*
b) i) *Maximum of 1 mark available.*
A large group is more efficient at finding food ***[1 mark]****.*
ii) *Maximum of 1 mark available, from any of the 2 points below.*
Grooming is hygienic ***[1 mark]****. Grooming helps to reinforce the social bonds within the group* ***[1 mark]****.*

Index

Index